AN AFRICAN EXPERIENCE

AN AFRICAN EXPERIENCE

GERALD HINDE • WILLIAM TAYLOR

SOUTHERN
BOOK PUBLISHERS

FOR MY WIFE PAM,
whose love and support contributed so much to this book.

Jesus said to him: " 'You shall love the Lord your God with all your
heart, with all your soul, and all your mind.'
This is the first and great commandment.
And second is like it: 'You shall love your neighbour as yourself.'
On these two commandments hang all the Law and Prophets."
MATTHEW 22: 37−40

Gerald Hinde

For two members of the next generation in the hope of a bright
future for Africa: Tully McClellan Cambell (19/4/93)
Alexandria Brittan Helena Filipowski (3/2/93)
And especially for Trisha with love

William Taylor

Copyright © 1993 by the authors

All rights reserved. No part of this publication may be reproduced or transmitted in any
form or by any means without prior permission from the publisher.

ISBN 1 86812 474 6
First edition, first impression 1993

Published by
Southern Book Publishers (Pty) Ltd
PO Box 3103, Halfway House, 1685

Cover design by Abdul Amien, Cape Town
Map drawn by Anne Westobe, Cape Town
Designed by Wim Reinders & Associates, Cape Town
Set in 12/14 pt Baskerville by 4-Ways DTP Services, Cape Town
Printed and bound by Singapore National Printers Ltd

CONTENTS

*A whitefaced owl in a
threatening pose.*

Page i: *The setting winter sun silhouettes the spiral horns of a kudu
bull looking back over his shoulder before heading for the safety of
the thick bush.*

Title page: *A zebra stallion, ever alert, keeps an eye out for danger
while his harem mates graze after a short afternoon thunderstorm.*

MANYOLETHI RIVER

MLOWATHI RIVER

TREKKER TRIALS CAMP

SAND RIVER

MATSHAPIRI RIVER

KRUGER NATIONAL PARK

HIPPO POOLS

MAIN CAMP

FLOCKFIELD BOMA

KAPEN RIVER

SAND RIVER

KIRKMAN'S CAMP

HARRY'S CAMP

MSUTHU RIVER

SABIE RIVER

KRUGER NATIONAL PARK

Mala Mala Game Reserve

FOREWORD

I am delighted that Gerald and William have honoured me by inviting me to write the foreword to *An African Experience*, especially as I believe that it is an excellent follow-up to Gerald's very successful book, *Leopard.*

I know that both young and old will greatly enjoy this African experience and will most certainly learn something about patterns of animal behaviour. It is also pleasing that Mala Mala has again been able to provide the base for Gerald's most professional photographic work and also for William's style of writing.

Mala Mala has always maintained that the level of human intrusion into virgin bushveld or any pristine area should be kept to the minimum to ensure the natural and ongoing renewal of the full spectrum of any particular ecosystem. This policy ensures that all elements sustained by and represented in a particular environment will survive for the benefit of posterity.

Fortunately, environmental conservation is a growing concern in many places around the world and Mala Mala's client entertainment programme effectively plays its part by introducing visitors from every part of the world to the many wonders of nature manifested in Africa.

I am sure that readers of *An African Experience* will thoroughly enjoy William Taylor's anecdotes and reports on his various experiences as a ranger at Mala Mala. I always knew that he was an excellent game ranger but I must confess that I was unaware of his journalistic talent.

William's anecdotes not only demonstrate some of the interesting wonders of the bushveld but also illustrate the enormous amount of daily interaction between the various creatures and plants.

In *An African Experience* Gerald and William have captured the essence of Mala Mala and the excellent opportunities it offers to enjoy and cherish unspoilt Africa. Photography fortunately creates sustainable trophies as the target may be captured, duplicated, stored and admired many times, rather than being a single forlorn trophy hanging on a wall.

Good luck to the new team of Gerald Hinde and William Taylor. May they continue to produce excellent works such as *An African Experience* for us and our children to enjoy.

Micheal L. P. Rattray

ACKNOWLEDGEMENTS

We wish to express our gratitude to Mike and Norma Rattray for granting us the opportunity to live and work at Mala Mala and for their generous support of this project. This is certainly one of the finest game viewing areas in Africa and combined with the excellent facilities and professional expertise of the Mala Mala staff, provides the ultimate African experience.

We would like to thank the management and staff of Mala Mala who were always ready with friendly assistance and expert advice. The catering staff provided us with excellent food and picnic baskets for the long vigils. The workshop manager and staff were always pleasant and extremely efficient at fixing vehicles and equipment at a moment's notice. Thanks to the staff of Harry's and Kirkman's camps who supplied us with petrol, food, drinks and friendship when working in the southern section of Mala Mala.

Thanks to Louise Grantham and the management and staff of Southern Book Publishers who have been a solid and helpful team to work with.

A big thank you to my family and wife, Pam, who have always supported, inspired and encouraged me in my work.

To all my friends, who have assisted us over the years, we salute you and thank you for the great work you are doing to promote conservation.

Many other people have helped us over the years − thank you.

Finally to the Lord God who has made all things possible for us: We give You all the glory. Let us remember to worship the Creator and not the creation.

Gerald

To individually acknowledge everyone who helped me would be impossible, so I would like to thank the many friends I made at Mala Mala, both staff and guests, all of whom helped make my three year stay enriching and enjoyable. To all my friends who stayed in touch and didn't forget me during my lengthy bush sojourns, thank you.

Thank you to my parents and family for continued support and understanding.

Thank you to Trisha Wilson for constantly making the seemingly impossible, possible, and for always being there.

William

Early morning dew glistens on the delicate strands of a spider web.

Identified by the ragged tear in his left ear, this enormous old bull elephant was a regular nocturnal visitor to the main camp, drawn by its succulent fruit trees.

Overleaf: *A young lion defends himself during a squabble at a kill.*

INTRODUCTION

Mala Mala is home to a wide variety of birds of prey.

Across the Sand River from the main camp at Mala Mala is a boulder strewn hill which rises from the banks of the Manyaleti River. This is Sitlhwayise, the highest point in the reserve, a mystic place of strange beauty, clothed in candelabra trees and haunted by the ghosts of Shangaan ancestors and old white hunters. If you climb to the top of Sitlhwayise early in the morning you can witness an event that never loses its magic – the sun rising over an unspoilt tract of African wilderness. The spreading light picks out pockets of mist in the dry riverbed which slowly disappear in the growing heat. Birdsong rises up from the stand of riverine bush at the base of the hill, the sound of impala ewes calling to their young drifts across from the open plains in the distance and one feels privileged to share in this age-old everyday awakening.

On my first visit to the upper boulders of Sitlhwayise I watched a pair of black-maned lions moving regally along the dry bed of the Manyolethi towards the great expanse of bushveld in the distance. In the skies above, vultures wheeled effortlessly on the first thermal of the day in their endless search for carrion. Looking out over the beauty of the bushveld in the soft morning light always reminds me that this is part of the small percentage of Africa's land surface that remains unspoilt.

Increasing pressure from the rapidly growing population and the attendant destruction of habitat has left only small pockets with viable ecosystems, and has caused a frighteningly rapid downward spiral in wildlife numbers. Only one per cent of the animals that lived in Africa 300 years ago remain. Not so long ago Mala Mala was also on its way to joining the shuffling queue of casualties; settlers on the central plateau of South Africa would descend on the lowveld in the winter months when the mosquitoes were inactive and meat could be dried in the cool winter air. They brought guns and supply wagons and set about decimating the seemingly limitless herds of animals. Others came bringing another agent of destruction – domestic livestock. Cattle and goats grazed on the lowveld grass-lands, trampling root stocks and changing the species mix of sensitive grass communities. Very soon these pressures manifested themselves in a drop in species numbers and in a visible impact on the environment: erosion ditches, bare earth, felled trees and trampled waterholes. It took a few men of vision and determination to stem and eventually turn the tide, to allow the land to slowly heal its wounds and regain its balance.

Although usually confined to thicket vegetation, bushbuck will sometimes feed on small herbaceous plants in forest clearings.

As you let your gaze wander westward from the top of Sitlhwayise, you can pick out the thatched rooftops of the camp amongst the marula and jackal berry trees on the bend of the river. This unobtrusive presence of man is a reversal of early times. Visitors now come to the Mala Mala Game Reserve to steal a glimpse of how things used to be, to delight in the excitement and the tranquillity that only the African bush can provide. The people who share this experience are reaping the benefits of an ongoing commitment to the conservation of an unspoilt wilderness.

As the sun rises higher, the hues change from soft pastels to harder, brighter reflections off rocks, red earth and green trees. Where the lions strode golden in the dawn a dusty grey elephant bull now moves slowly towards the glistening waters of the Sand River. My companions on the rock are a broadcasting executive from Johannesburg and his wife, a satellite technician from London and an artist from Dallas. Although they come from different backgrounds and all the corners of the earth, they are all enthralled by the simple but spellbinding show below us.

From Sithlwayise we make our way through Paraffin Drift along the road that twists and turns on the banks of the Sand River, passing through the huge trees and thick undergrowth of the riverine forest. This rich plant community supports an array of browsers and fruit eaters; bushbuck, kudu, impala and duiker feed off the leaves of vines, bushes and trees. Monkeys, baboons and squirrels take advantage of fruiting fig trees, as do many of the 300 or more species of bird found in the area. There is always a chance here of coming across that elusive and most beautiful of the big cats, the leopard. We continue our journey slowly, carefully scanning the ground and thickets for signs of a female leopard who has been seen here with two cubs.

Having learnt how to navigate the 45 000 acres of pristine bushveld in the reserve, one is able to concentrate on locating and identifying animals. Eventually one can recognise individuals and by pinpointing the various sightings soon get a fair idea of territorial boundaries.

The road eventually brings us to the seclusion of the hippo pools. A large sycamore fig tree provides the shade from where these great beasts can be viewed lazing the daylight hours away in the limpid pools formed by natural rock walls across the river. Every so often the quiet of the morning is shattered by a loud grunting and a shower of spray as a hippo surfaces to breathe. The call is picked up by the rest of the herd and continues for some time before peace again settles on the pools.

I (William Taylor) worked as a game ranger at Mala Mala for three years during which time Gerald Hinde was completing photographic work in the reserve for his book on the leopards of the area. After spending many hours in the bush together the idea of this book began to take shape: a book which describes the stories and experiences of the Mala Mala Game Reserve, illustrated with Gerald's photographs of its inhabitants, large and small. But more than just Mala Mala's story — a story of Africa and her creatures, a re-creation of our experience of Africa. Encouraged by Michael Rattray, the owner of Mala Mala, we put the book together. The result is a mixture of anec-

Threebanded plovers hunt actively along temporary pools and rivers.

Male lion partnerships are close associations; constant tactile communication is an important part of the continued bonding process.

dotes from years of following animals, in particular the big cats, information on the ecosystems and animals of the bushveld and a dramatic photographic record of games drives in unspoilt Africa.

The continuing success of reserves like Mala Mala is essential to the maintenance of wildlife stability in Africa, its success has afforded visitors from around the globe an experience of Africa at its best. We hope in some small way to have captured the magic, the power and the peace of Africa in these pages.

Mala Mala main camp blends into the bushveld on a bend in the Sand River.

A place of great beauty.

The beady orange eye of a Cape glossy starling scans the ground for insect movement.

Overleaf: Two young wildebeest bulls chase one another through lush summer grass.

5

HOW IT ALL BEGAN

Two members of the Styx Pride fix a spine-chilling glare on a minor source of irritation.

"Ngonyama," he whispered as his hand shot up indicating to the hunting party that they should be more cautious. The spoor they had been following from the wildebeest crossing on the Sand River was almost the size of a dinner plate. They were on the track of a male lion. The rustle of dry leaves underfoot alerted the two male lions lying in a thicket west of the thin dusty track that ran from Mala Mala to Skukuza camp in the Kruger National Park. The lions lifted their massive shaggy heads and, after a brief stare, moved off into the thick bush. A number of tawny heads popped up from the grass 30 metres further south. The hunting party remained dead still as the pride tried to locate the cause for concern. Three large subadult males slowly relaxed. They stood up, stretched lazily and then lay down a pace or two from their original position. One flopped onto his side − the other two settled on their bellies and began grooming themselves, looking up occasionally as if still slightly concerned. The presence of lions and the aura of power that surrounds them sent a chill down the spines of everyone in the hunting party. The size and strength of these, the largest of Africa's great cats, have always inspired awe in man and perhaps it is the challenge of overcoming this fear and competing with a fellow predator that has drawn people to hunt lions.

When the pride relaxed Tayi Mhlabo, the Shangaan tracker, beckoned to the lady who was to do the shooting, indicating that she should move slowly forward over the few yards that separated them. When she was level with him, Tayi pointed to the young male with a slightly fuller mane than the others. She raised the heavy calibre rifle to her shoulder, aimed, took a deep breath, then gently squeezed the trigger. The loud report sent the pride scattering in all directions − no one from the hunting party moved. Tayi was the first to show any emotion, a broad grin revealed a gap in his upper row of teeth. "Ufile," he announced emphatically, "He is dead!" The great monarch breathed its last as the parched soil drank at the trickle of blood that flowed from the base of its skull.

To this day the area is known as "Princess Alice's bush", the place where Princess Alice, the lady in the hunting party, killed her first and only lion. The sound of shutters releasing and motor drives winding on have replaced the loud reports of rifles. Images on negative and positive film are the trophies and the animals are now left for others to appreciate and photograph.

A good long drink after a dry and dusty winter's day.

9

The Eastern Transvaal lowveld of South Africa has long been known to man as an area of plenty. In rocky outcrops of the Kruger National Park testimony to the abundance of wildlife has been left by the San bushmen who hunted and gathered food in this area for thousands of years. Their art, painted delicately onto caves and over-hanging rocks with paint made from lichen and bark, bring their hunts to life – impala leap out from the rocks and rhinoceros amble across the cave walls. The artists are gone now but the animals are still here, as they have been for millions of years.

The lowveld is bounded by the Crocodile River in the south, the mighty Limpopo in the north, by the escarpment of the Drakensberg mountains in the west and the Lebombo range in the east. It is the extension of an ecosystem that runs the whole eastern length of Africa. From the savannas of the Mara and Serengeti it runs south-wards along the Great Rift Valley with its deep lakes, through Zim-babwe and the lowveld and into Swaziland and Zululand in the south. A staggering diversity of game species lives in this system – the biggest of all land animals, the teeming herds of plains game and the most savage of predators.

In the heart of the Transvaal lowveld lies the Sabie Sand Game Reserve and within this reserve is the gem of the bushveld – Mala Mala.

Nature has held sway here as sole ruler for millennia, until recent-ly the ravages of malaria and sleeping sickness, transmitted by the bites of tsetse flies, were her defence against man and his livestock. But slowly man pushed deeper and deeper into the lowveld. The first settlers after the San were probably black people – migrants from the north who lived in relative harmony with the habitat. Game was plentiful, the land was fertile and clean water flowed in the rivers running through the bushveld to the Indian Ocean in the east. These men must have experienced the brooding beauty of the lowveld with the roar of the lion and the cough of the leopard to remind them that, although they were free to explore during the day, the night belonged to others.

The only contact these people had with the disturbing influences of the outside world was probably the occasional encounter with Portuguese explorers in the early eighteenth century who recorded meeting friendly black tribes in the area. But a more sinister and powerful influence was to come from the south, from people who lived in the rolling hill country on the east coast – the Zulu. In 1817 a man of vision and great cruelty succeeded Dingiswayo as chief of the Zulu tribe. He was a military genius whose aim was to conquer, subdue and crush the tribes around him, his name was Shaka and he was to change the face of warfare in southern Africa. Shaka moulded his impis or regiments into fit, highly skilled fighters. He cut the shafts of their spears down to handle size, lengthened and broadened the blade and created a short stabbing spear – the assegai. This fear-some weapon never left the hand of the Zulu warrior. Shaka taught hand-to-hand combat, co-ordinated attack formations, chose leaders with care and weeded out slackers with impunity. He then went to war beginning an era of bloodshed that sent shockwaves rippling through Africa. This was the time of the Mfecane, the crushing.

One of Shaka's generals, So-Shangane, arrived in the Mala Mala area with an impi of 800 men; he rampaged through the lowveld,

Contrary to popular belief, leopard spend more time resting on the ground than in trees.

laying everything waste – conquering the black tribes and sending Portuguese settlers scuttling for safety. The conquered people, the Bathanga, were assimilated into So-Shangane's Gaza empire where their language and customs influenced the Zulu way of life. The people that emerged from this mixture derived their name from that of their leader, and today the Shangaans are still the dominant black tribe in the lowveld.

The first of the Dutch settlers, who had been established in the Cape since 1652, started to arrive in the lowveld in the 1720s. They were after gold, ivory and the fabled kingdom of Monomatapa which they believed to exist in the central part of the subcontinent. In 1772, after taking Delagoa Bay, on what is now the Mozambique coast, they sent an expeditionary force westwards from the coast. The leader of this small band, De Cuiper, kept a meticulous record of his journey, and on 10 June 1725, his diary says they crossed the Crocodile River stepping into land that would later become the Kruger National Park. They only got as far as the Sabie River before they were attacked by a hostile tribe and had to retreat.

No other white people arrived in this area until 100 years later when Captain Cornwallis Harris of the East India Company came to hunt and explore in 1836. He was a sportsman and a fine artist who recorded his travels on paper and canvas. He was also a meticulous observer of nature, particularly of the animals he hunted, and it is to him that the discovery of the Sable antelope, the emblem of Mala Mala, is accredited.

By 1836 there were other white men in the lowveld. Intrepid and hardy Boer settlers were seeking out pastures new, away from British domination in the Cape. Trichardt and Van Rensburg – two Boer leaders – pushed north through the Transvaal, Van Rensburg headed east for the Mozambique coast but his entire party was wiped out by So-Shangane's men. Louis Trichardt followed but his cattle were killed by lions and bitten by ticks and his people succumbed to malaria and sleeping sickness. He finally reached the coast in 1838 with only 26 survivors.

Hunters fear no animal as much as a buffalo bull in a reedbed.

More and more Boer settlers moved to the Transvaal were they settled on the eastern edge of the plateau above the lowveld and established the Ohrigstad Republic. Many of these men would go down into the lowveld with wagons during winter when mosquitoes and the diseases they bore were inactive. They began systematically to slaughter the herds of game for meat and hides. The only white person living in the lowveld at this time was a Portuguese man, Joao Albasini. He lived on the Sabie River where he hunted ivory and meat for trade with Portugal. Joao eventually married the daughter of one of the Boer trekkers and left his lowveld kingdom in 1847. In the early 1850s the authorities of the Ohrigstad Republic began to notice a drastic drop in the numbers of game in the lowveld. In 1858 a resolution was passed limiting the amount of game that could be shot; only one wagon per person could be loaded with meat and hides and no animals were to be shot for hides alone. This law itself shows how devastating and wanton the slaughter must have been. As it was the law was never enforced and the destruction continued.

It was in the 1860s that the records first mention the farm Mala Mala. Its emergence is linked with a name that has since become synonymous with conservation in Africa, Paul Kruger. The young

Lion cubs will often gorge themselves until the effort of staggering full bellied from a carcass is altogether too much.

Kruger led a party of men, mostly foreigners from gold diggings in the area, to quell an uprising of blacks in the Northern Transvaal. John McNab, a member of this party, was bestowed the farm Mala Mala, at a charge of £1.10s, for services rendered. He was granted full title to the land, for which he had to pay a sum of money annually, as well as make himself available for military call up at any time. He had to allow travellers to water and feed their stock and he had to keep the roads and paths in good order. In 1893 the farm was bought outright by John Wylie Craig Niven and he had it officially surveyed.

In March 1898 Paul Kruger's dream of establishing a sanctuary in the lowveld finally became a reality with the proclamation of the Sabie Game Reserve. This consisted of the low lying areas between the Crocodile and Sabie rivers, 4 660 square kilometres of land. Three factors influenced this historical event: the miraculous disappearance of the tsetse fly after the rinderpest epidemic in 1896, the new gold money in the coffers of the Transvaal republic and the tenacity of Paul Kruger. It was also at this time that Kruger started work on the Selati railway line linking the goldfields in the east with the Pretoria-Delagoa Bay line. The stone ballast of this line can still be seen running through the southern part of the Mala Mala Game Reserve.

The history of the area became increasingly eventful. Farms were

The Borehole Pride lie up in the Sand River under a threatening sky.

bought by the Transvaal Consolidated Land and Exploration Company and other large organisations. The result was that government and private land was mixed. Colonel James Stevenson-Hamilton, who was the first warden of the Sabie Game Reserve, persuaded landowners to hand their farms over to the government for five years in exchange for services which the staff of the reserve would render. The boundaries of the reserve were extended twelve miles west and in this way Mala Mala fell within the Sabie Game Reserve. The farm Mala Mala and a number of surrounding properties were bought in 1921 and cattle farming began − much to the chagrin of Stevenson-Hamilton. After much in-fighting and a court case involving the shooting of an wildebeest bull by the farm manager, the government finally resolved the issue by moving their fence eastward − Mala Mala was again outside the reserve.

In May 1921, Mala Mala was bought by William Alfred Campbell, known to all as "Wac". Wac Campbell brought many colleagues and staff from his Natal sugar estates on hunting trips to the farm. He wrote a letter to the Transvaal Provincial Secretary stating his objectives for Mala Mala: to conserve fauna and flora and to enforce the strictest preservation of game with the objective of passing the farm on as a legacy to his children. Wac, his friends and employees continued to hunt and enjoy winters at Mala Mala until 1964 when the farm was sold by Wac's son to Loring Rattray. This change of ownership heralded the beginning of a new era in Mala Mala's history.

Throughout its history, Mala Mala had been part of an area where the gun had a heavy influence, wild animals were either competition for cattle or potential trophies for hunters. During Wac Campbell's tenure as owner the emphasis slowly began to shift away from hunting, it was becoming a place where people could get away from their everyday lives and relax in the bush.

Loring Rattray already owned the farms Exeter and Wallingford in the Sabie Sand Game Reserve, and in 1964 the prize of them all was his. The operation which Wac's son, Urban Campbell, had begun was extended and refined and soon became locally known as a quaint and peaceful lodge to relax and view game.

Loring died in February 1975 at the age of seventy-one, leaving his estate to his wife Natalie. Their son Michael acquired the Mala Mala business and approximately 7 000 acres from his father's estate with borrowed capital. Michael was in his early forties and had made a reputation for himself as a successful farmer and businessman in Natal. He was very interested in conservation and determined to make Mala Mala a commercial success. When he bought it, it was running at a financial loss.

By introducing solid conservation principles and practices Michael Rattray slowly restored the natural balance of the reserve and with the cessation of hunting, game viewing improved dramatically. Once Rattray had acquired the properties he needed to establish a game reserve that would function as an ecologically viable unit, he set about laying the plans for its operation. In order to maintain constantly good game viewing the running of any conservation area must be based on a sound policy of management. Mala Mala had adopted a philosophy of wildlife management which was developed into a policy in the early 1960s. The concept of a game reserve that could be run purely for the enjoyment of viewing and photographing rather than hunting was a novel one and not generally accepted as viable.

The first consideration must always be financial. How does one generate enough money to run a management policy and maintain a large area of land without interfering with the natural systems? From the early days it was decided to keep the numbers of people visiting at any given time down to a minimum, which made Mala Mala exclusive and more expensive, but helped maintain the environmental integrity of the reserve and reduced human impact on the ecosystem.

Mala Mala now covers 45 000 acres of bushveld wilderness and is the finest private game reserve in the world. The veld and game management programmes maintain some of the best game viewing on the continent allowing visitors to relax and experience Africa at its best. Mala Mala's popularity has risen dramatically.

A ground hornbill moves through the grassland in search of prey. These large birds spend most of their time in small groups on the ground where they prey on insects, lizards, snakes and small mammals. Overleaf: Impala lambs are all born in a short space of time in spring and early summer. During the day they are left together in nurseries while their mothers feed and keep an eye out for predators.

Female hyenas form highly social clans in which they scavenge, hunt and raise young together. Each mature female gives birth to two tiny black pups which are raised communally.

Play and mischief are elevated to an art form by the young members of a bushveld baboon troop.

The greyheaded bush shrike is one of the larger and more powerful of this magnificently coloured and liquid-voiced family of forest birds.

An African jacana heads home after spending a day gleaning insects off rafts of vegetation at the hippo pools.

Short white manes erected to make themselves look bigger, two kudu bulls spar vigorously at the beginning of the mating season. The shape of the horns sometimes leads to entanglement and a slow death.

The African elephant is a relic of one of the most successful groups of large land mammals ever to walk the earth, a group known as the proboscideans which includes mastodons and mammoths. They can subsist in almost any habitat and are found from the thick rain forests of West Africa to the desert of Namibia's Kaokoveld. Their preferred habitat is an area with both grazing and browsing, ideally a forest edge with a variety of trees, shrubs and vines and a plentiful water supply. They spend about 16 hours a day feeding their vast bulk and the remainder of the day resting, bathing or wallowing.

The social organisation of elephants is as complex as any other social mammal. Although the Sand River is more heavily utilised as a "retirement" area by big old elephant bulls, breeding herds do pass through, affording a glimpse of this fascinating society. The basic unit of the elephant herd is a group of related females led by the matriarch, always the oldest cow and usually the mother of two or three of the adult females. Typically this group numbers between 8 and 12 animals.

The matriarch is a wise old animal who initiates all movement, determines feeding grounds and around whom the whole herd clusters when disturbed. Unlike other mammals the importance of experience and leadership is so great in the society that females continue to lead herds long after they have ceased to be reproductively active. It has often been observed that the sudden death of a matriarch precipitates the herd into chaos and frenzy. Some of the older females will try to lift her to her feet with trunks and tusks, squealing and trumpeting, and if they do manage to get her up they press close in on her sides to support her and move her away. Their distress is obvious and the altruism in the society so well developed that the whole herd will mill around and expose themselves to danger rather than run away.

When family groups grow too large they usually split but still greet each other with excitement and obvious recognition when they meet at waterholes or feeding grounds. Recent research has revealed a remarkable range of infrasonic vocal communication

amongst elephants. It has long been known that the "stomach rumbles" of elephants in close contact with one another are a form of communication but it seems that this is only the upper range of a whole vocabulary of long wavelength, low frequency sounds too deep for detection by the human ear. The nature of these soundwaves allows them to carry for up to eight kilometres through thick vegetation and keeps individuals and herds in constant contact with each other. It allows females to avoid sexually active males who become aggressive when in a condition of peak hormonal activity known as musth. Bulls in this state can be recognised by copious secretions from the temporal glands behind the eyes and a foul smelling secretion from the swollen genital area. It is wise to avoid a musth bull. When elephants meet they engage in an endearing greeting ceremony, touching and twining trunks and then placing their trunk in the other's mouth.

Southern Africa has a healthy, growing population of elephants but as with elephants throughout Africa the danger of poaching for ivory is a constant threat. In the Kruger National Park and parks in Zimbabwe, culling of elephants has become a necessary management practice in order to curb the destructive influence large populations have on the habitat. Unfortunately these huge gentle, creatures can no longer spread themselves evenly throughout their range but are confined to pockets of habitat in game reserves. Culling of entire herds is the most humane way to control the elephant population and the operation, which removes roughly 700 animals a year from the Kruger National Park, is swift and merciful. All elephant carcasses are processed and the meat and hides are sold to generate money for the perpetuation of the habitat, ensuring the continuing survival of the species.

A PLACE OF GREAT BEAUTY

Little bee-eaters are often seen hawking hatching insects from reeds along the river banks.

Giraffes must be totally sure that no predators lurk nearby before adopting their vulnerable, spread-legged drinking position.

Bushveld is a word that describes an area containing a myriad of habitats. Some areas are totally dominated by mopane scrub veld for hundreds of square kilometres with skylines dominated by the massive baobab tree, some are mostly grass and others are dominated by sandveld tree communities. Mala Mala cannot be categorised by the most dominant habitat type, its strength is habitat diversity.

The reserve covers 45 000 acres of gently rolling country more or less bisected by the perennial Sand River, the most important source of water in the Sabie Sand Game Reserve. The average precipitation per annum is 700 to 750 millimetres, with rain falling almost exclusively during the summer months. The rain usually starts in October with thunderstorms that build up hot and heavy during the day and lash the countryside in the afternoon and evening. During the spring months lightning sometimes sets the tinder dry bush alight and fires light up the night sky. By April the rain has stopped and the following months are dry as the season moves into the clean blue daytime skies and crisp, star laden nights of winter. The rain however is fickle and drought is not unusual.

Although the communities of plants, the soil types and landscapes are extremely varied, a few major habitat types are distinguishable. Along the banks of the river is a dense forest dominated by jackal berry trees with a thick understorey of bushes, scrambling shrubs and vines.

Away from the river the country falls away in a series of low, rolling hills. On the tops of these ridges the vegetation is dominated by a mixture of bushwillow, knobthorns and marula trees which form an open woodland with long rank grasses between stands of trees.

The valleys in-between these ridges support delicate grassland communities, based on clay soils, that are known as seeplines. These areas are very important to the large grazing species such as rhino, wildebeest, buffalo and zebra who feed on the red grass, buffalo grass and other nutritious grass species that thrive in the wet summer months. Some trees grow on the grassland, usually large and well-spaced individuals such as the leadwood, knobthorn and other *Acacia* species.

In the south of the reserve there are areas of abundant bush with dense stands of the resilient *Strychnos* bush on the ridges and a mixture of many species including buffalo thorn, bushwillow, raisin bush and silver cluster leaf trees, on the slopes leading down to the drainage lines.

A regular bath keeps off irritating parasites, cools down the body and provides an opportunity to top up with water.

Throughout the reserve there are intermittent rocky outcrops and some large granite *koppies* or hills. These are an important component of the landscape – not only for their different vegetation types but also for the cover and vantage points they provide. Drainage lines or *dongas* that are formed by rain run off into the Sand River whose tributaries are another major relief feature that provides excellent game viewing as they are used for travel and feeding by most game species.

This great diversity of habitat supports a wide range of animal and bird species, from the tiny but ubiquitous termite to the majestic elephant, from the busy nectar-feeding sunbirds to the ostrich of the grasslands. Insects and reptiles abound and it is as fascinating to

Early morning mists make for spectacular winter sunrises.

watch a dung beetle roll his mate along on a ball of fresh dung, or a black mamba exploring nests and holes in a leadwood tree, as it is to watch a lion stalking or two buffalo bulls duelling over a female.

In the summer the bushveld becomes a lush blanket of green spreading as far as the eye can see. When autumn comes the trees begin to change colour and beautiful vistas of rolling hills clothed in variations of rust, orange, red and brown delight the eye. In winter the green-clad banks of the Sand River stand out from the surrounding grey countryside and the dusty air and clear skies combine to produce the most spectacular of African sunsets. The diversity of habitat and everchanging cycle of seasons create an experience of unending fascination.

27

Scarlet-chested (top) and collared sunbirds feed on the nectar of the lucky bean tree, Erythrina caffra.
Left: A klipspringer perched atop his rocky kingdom at the summit of Sitlhwayise.
Overleaf: A large herd of buffalo race down to a dwindling pool in the Sand River. During winter long trips from feeding grounds to water must be made daily. The proximity of water often causes a stampede.

Pseudocleobotra wahlbergi.

After the uniform brown and grey of the bushveld in the flat light of the winter months, the rains bring on a burst of colour almost overnight. As animals that can discern colour and that have evolved a complex sense of aesthetic appreciation, we are often tempted to think that this glorious emergence of grasses, flowers, insects and birds is especially laid on for the purpose of pleasing our eye. But colour in nature means more than just beauty.

After the first rains the riverbeds and grassy plains are dotted with flowers. There is the lion's eye, a single bright orange flower borne on a long thin stem, which grows amongst the grass in disturbed areas. In the riverbeds a tiny purple and yellow flower, *Cleome maculata*, presents a vivid pattern to bees until the flower has been pollinated and then the lower petal droops, hiding the pattern and encouraging insects to visit other flowers on the bush. The wild hibiscus, wild foxglove, the morning glory, Transvaal sesame and tree wisteria, all cousins of popular garden plants, use bright colour to attract insects. The bright colour

alerts insects to the presence of food, usually a nectar, rich in natural sugar and energy. Each species of plant has adapted its structure to its pollinator, bee, fly, butterfly, beetle or moth, in such a way that to reach its reward the pollinator must first become dusted with pollen. When visiting the next flower the pollen is rubbed off and fertilisation takes place.

Insects themselves show a myriad of colours of various functions. The African monarch, a butterfly, is an example of aposematic colouration, its vivid patterning foregoes camouflage to warn predators away. The African monarch feeds on a poisonous plant of the milkweed family and stores the toxins in glands in its abdomen. When eaten these toxins are released and bring a foul taste to the predator's mouth. Over time birds have learned to avoid the monarch and it is seldom attacked. Other species of butterfly have cashed in on this and copy the pattern of the monarch so closely that it is difficult to detect them as fakes. This behaviour, known as mimicry, occurs in many insect families including a type of wasp found in ants' nests in the

bushveld. The wasp has lost its wings and taken on the form of the ants it lives amongst and preys on.

A beautiful praying mantis, *Pseudocleobotra wahlbergi*, with complex structural patterning is found at Mala Mala. When stationary on a branch it is almost impossible to detect. It waits until an unsuspecting insect wanders within reach and then lashes out with its raptorial forelegs, grasping its prey and settling back to feed on the body fluids.

The most conspicuous colours amongst the trees and bushes are the flashes of birds in bright breeding plumage. In the spring the bird population grows by almost 40 per cent as the migratory birds arrive from the north. The woodland kingfisher's presence is announced in mid-November with its trilling call and flashes of bright turquoise as it darts from tree to tree in the hope of attracting a mate. Some male birds that spend most of their year in drab plumage suddenly moult into bright new colours in the summer. The male red bishop comes into breeding plumage in November and can be seen flying around like a little red general attended by a flock of his dull brown females. The brighter his plumage and the more vigorous his display, the more females he will attract. A bird's colour is a reflection of its condition, and its condition is a reflection of how good the food resources are in its jealously guarded territory. The wise female does not choose a mate because his brighter plumage and more vigorous display make him dashing and handsome, but because her chances of raising her young successfully in his territory are good, and his genes for strong competition will be good for her offspring.

Colour is also important to mammals that are thought not to discern colour well. The higher primates have a very important behavioural response to the colour of their young. Vervet monkeys and baboons are both born with black fur and pink faces − this colour combination triggers a powerful maternal response in females. When young vervets are in this pelage phase they are handled almost continually by juvenile females in the troop with the consent of the mother; young baboons too are inspected constantly by subadult females in the troop until their colour changes and they are left alone. Amongst the ungulates, both buffalo and wildebeest have lightly coloured young that encourage a strong protective behaviour from the rest of the herd until they gradually assume adult coloration.

Perhaps the most comical of the messages sent using colour is the "flagging" behaviour of the blue headed agama lizard. This large lizard lives clinging to the trunks of old trees. The males have a bright blue head which they bob up and down in a territorial threat display to other males. With their small eyes and loose jowls they look like old blue headed men nodding in time to a catchy tune.

Blue headed agama lizard.

THE RIVER

The plump green pigeon is a resident of the riverine forest canopy.

A member of the Styx Pride tests the breeze and follows the aerial progress of a pair of vultures while emerging from the reedbeds.

The Sand River is a vein of precious liquid running roughly north–south along the length of Mala Mala. It originates on the highveld escarpment hundreds of kilometres north-west of the reserve and flows through the homeland of Gazankulu and the lowveld farming areas, before entering the greater Kruger National Park conservation area. The water that it brings is the lifeblood of the game reserve – in the dry winter months game must migrate daily to the river from grazing areas on the edge of the river valley.

The riverine forest, a thicket along the banks of the Sand and its larger tributaries, is one of the more important and certainly the most productive of the habitats in Mala Mala in terms of species diversity. Because of the abundance of underground water, huge trees, typical of riverine forests, line the waterways; jackal berry, sycamore fig, matumi and weeping boer-bean are amongst the more common. Many of these and some of the smaller plants in this habitat produce fruits and berries throughout the year which are an important food source for frugivorous birds.

There is a huge sycamore fig tree on the banks of the river upstream from the main camp at an area known as the hippo pools. I have spent many hours daydreaming under the spreading branches of this giant, watching the parade of avian splendour that passes through the canopy and allowing my mind to follow it.

Probably the most beautiful and conspicuous of the birds in this habitat are the purple crested louries – large birds that hop clumsily amongst the branches and leaves plucking off fruit with stout beaks. At rest and without the sun on them, louries can appear quite ordinary. But the sunlight picks out a sheen of greens and purple, and when a lourie spreads its wings to fly, the brilliant red flash of the primary feathers is breathtaking. A copper-based organic pigment called turacin, which derives its name from that of the lourie genus *Tuaraco*, is the chemical responsible for this extraordinary colour. The feathers are highly prized throughout Africa and South America for ceremonial costumes and headdresses, which has contributed to a drop in population numbers of these birds in some areas.

A fellow frugivore is the green pigeon, a brightly coloured comical-looking dove which clambers around the trees while feeding, like a parrot. They keep up a low, purring chatter to communicate with one another until they are disturbed. A sudden silence is often a good clue to the proximity of a snake, eagle or large predator.

Vervet monkeys are another species that takes full advantage of the nutritive value of the fruit and flowers of the trees along the river and in so doing open a new feeding niche for some of the riverine antelope, in particular the bushbuck. The relationship between the monkeys and bushbuck has not been studied scientifically but observations indicate a definite link. It seems that wherever a troop of monkeys is feeding in the canopy, a small group of bushbuck are in attendance on the forest floor below. The bushbuck derive two major benefits from this. Firstly, when vervets feed they often peel or pluck choice pieces off their meal and drop the rest – which they are able to do with their dextrous hands. The discarded portions are high in protein and represent a valuable resource to the bushbuck waiting below.

In addition the monkeys are extremely vigilant and normally have sentries posted to warn the troop against the presence of predators. Their primary enemy is the leopard, and the monkeys with their keen eyesight and elevated position very seldom let a leopard sneak by undetected. They have an alarm call reserved especially for leopards, which are the bushbuck's major predator. A chattering cackle interspersed with high-pitched squeaks immediately alerts other animals in the area, including rangers and trackers searching the bush for that elusive prize among the predators, to a leopard's presence. Game rangers throughout Africa pay special heed to the alarm calls of animals, from squirrels to elephants, and usually when hot on the spoor of the big cats, it is an alarm call that will lead you to your quarry.

I was interested to note that rangers working on a research project on the tigers of India used many of the same techniques when tracking the big cats. The fauna of Asia Minor and Africa are almost identical: they have leopard, lion, rhino, elephant, antelope and the langur monkey, very similar to our vervet, and smaller predators like the genet and civet. This is because India and Africa were originally part of the same supercontinent known as Gondwanaland. Millions of years ago, Asia Minor separated from Africa and drifted on a large plate toward Asia leaving Madagascar in its wake. It moved up against Asia pushing up the Himalayan range as it did so. The animals on this huge moving continent evolved on its many-million-year journey and became geographically isolated between the Himalayas and the ocean. Much of their behaviour remained the same during that process: the langur monkey in India is as good an ally in predator detection to Indian game rangers as vervets are to rangers in Africa.

Thoughts drift away from the canopy of the fig tree when a hippo breaks the surface of the water with a loud exhalation of breath and a long sigh, drawing attention to the river itself. One often forgets in this dry land that the river is not only a source of water but also an extremely important food source, a habitat that supports its own unique ecosystem. The Sand River is classified zoogeographically in the afrotropical faunal zone and has a different fish fauna from more southern rivers. Species found in the Zambezi, Kafue and Limpopo rivers to the north also occur here, most notably the tiger fish, many tilapia and bream species and catfish or barbel. Barbel, the main

A huge crocodile launches itself into the waters of the Sand River after being startled by our helicopter.

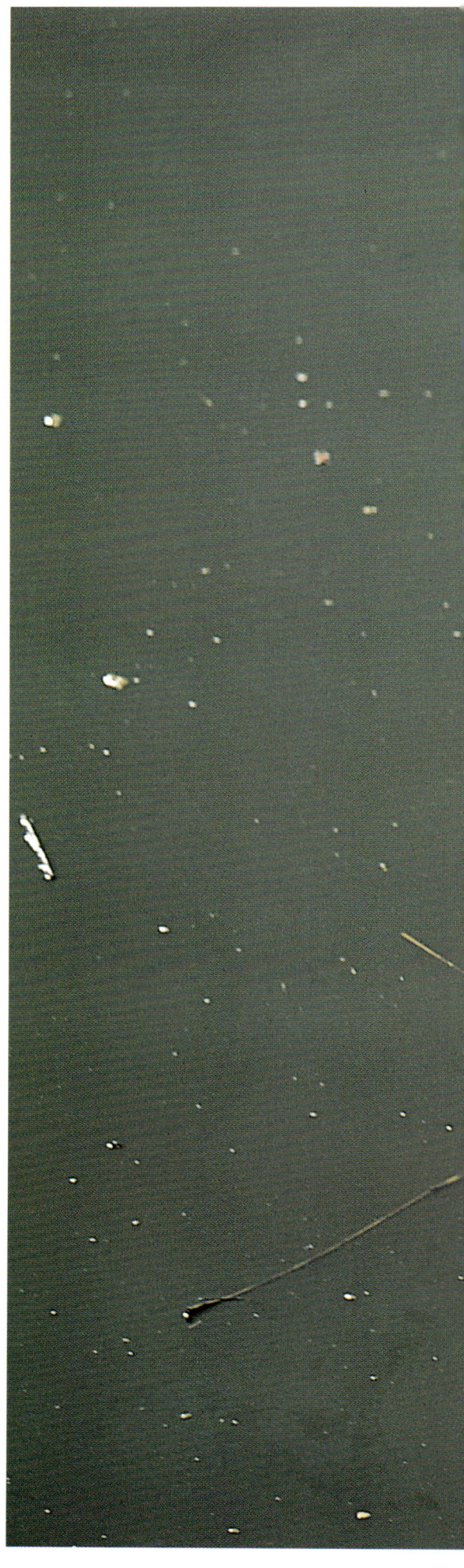

36

source of food for crocodiles in African river systems, are ugly, powerful bottom feeding fish that eat anything they can fit into their cavernous mouths. They can weigh as much as 22 to 27 kg. Crocodiles feed heavily on these fish particularly when they become concentrated in river systems for breeding. The other major predators of fish in the Sand River are the guild of piscivorous bird species – eagles, cormorants, herons and kingfishers in particular.

The beautiful chestnut and white African fish eagle is a symbol of the waterways of the continent, its wild ringing cry is distilled Africa. Along the Sand River, visiting fish eagles can be seen perched, waiting to swoop down and snatch an unwary fish from just beneath the surface of the water.

Along the stretches where the river runs shallow over rocks or sand, one can hope to see the stately figure of the goliath heron. I once watched one of these beautiful birds fishing south of the camp, staring intently along his beak waiting for the right fish before striking with lightning speed and perfectly skewering a large tilapia. Studies have revealed that goliaths often make only one strike every hour. Apparently they prefer one large meal to a number of small ones. They usually skewer their prey, unlike most similar birds which grab prey in their beaks. An interesting encounter between a goliath heron and a fish eagle was reported in the British ornithological journal, *The Auk*. A goliath heron had caught a fish and was about to swallow it when a pirating fish eagle swooped down to steal it. Within seconds both birds were dead. The heron had run the fish eagle straight through with its beak while the eagle's talons were clamped around the heron's neck.

Perhaps the most beautiful of all the birds that feed, swim, wade or drink along the river is the saddlebilled stork. They are usually seen in pairs wading with slow deliberate strides through the water, occasionally breaking into an awkward long-legged sprint, with wings held out to the sides in pursuit of prey or merely to impress a mate.

Moving out of the shady trees on the riverbank onto the sandy bottom of the riverbed reveals a different habitat. The riverbed is vegetated mostly by reeds of the genus *Phragmites* which grow in thick stands along the watercourse. Wild date palms and river bushwillows grow scattered in the open sand, and in some places where the river forms pools, bulrushes and water lilies grow. In the larger pools the Sand River supports healthy populations of hippo and crocodile, although the sinister form of the latter is likely to be seen anywhere in the river, often in the most unexpected places.

Testimony to the lurking presence of crocodiles is the story of the death of the last black rhino in Mala Mala in February 1991. Crocodiles are very seldom seen on game drives and guests will often express doubt about the existence of any crocodiles in a river that sometimes shrinks to a mere trickle, but when there is food to be had crocodiles appear as if by magic.

On this occasion an old black rhino bull which had been seen fairly regularly was found dead, apparently from a wound received in a fight with a white rhino, some 20 metres from the water. That night the river began to rise, swollen by heavy rain in the catchment areas upstream. Early the next morning the rhino was being fed upon by a pair of young male lions, in the evening we watched fascinated

Top: *A purplecrested lourie interrupts its feeding to give a raucous contact call from the canopy of a riverine tree.*
Bottom: *The vivid scarlet flash of a lourie's primary flight feathers, displayed by the very similar Knysna lourie.*

The enormous wingspan of a saddlebilled stork.

as a large crocodile appeared out of the water and took over the carcass from the lions. Within an hour the water was lapping up against the rhino carcass and it was being fed on by seven crocodiles all over 10 feet long! The lesson in this story is obvious – never take any stretch of water in Africa for granted.

The crocodile is an extremely dangerous predator. The fact that it has survived in its present form with very little change since the age of dinosaurs, 65 million years ago, is witness to its success. It was not until fairly recently that the fascinating story of maternal protection in the crocodile was revealed and it is worth having a closer look at its life cycle. After an elaborate courtship, mating takes place in July or August and by November the female is ready to lay her eggs. She comes out onto a sand bank, usually in the late afternoon or at night, and digs a funnel-shaped hole with her hind legs in which she carefully lays between 20 and 80 eggs. The eggs are covered with sand and the female moves off a little distance to watch over the nest. During this time she does not eat and actively chases off potential nest robbers. In the hole a miracle of nature takes place. The funnel shape causes eggs at different depths to incubate at different temperatures, and the position and temperature of the egg determines the sex of the hatchling. Eggs incubated between 27 and 29 degrees centigrade will only produce females and eggs incubated between 31 and 34 degrees only produce males. When they are ready to hatch the youngsters start emitting high-pitched yelps from within their eggs, a signal for their mother to uncover them. As they hatch she

puts them gently in her huge mouth and in several journeys takes them to the river where she releases them into a nursery area. Only after eight weeks does she eventually slide away and leave her youngsters to disperse and face life's perils on their own.

At the height of the rainy season, the Sand River is a raging torrent swollen above its banks. Six months before it was almost completely dry (right).

The river changes character and mood throughout the year and with the seasons. In summer the main watercourse usually flows strongly and occasionally after heavy rains the river will rise to a raging torrent, breaking its banks and hurling water down to its junction with the Sabie River near Skukuza. At this time of year the river is not an important source of drinking water for game, which has migrated out onto the open plains where rainwater has filled temporary pools and dams close to the grazing areas. It is in the dry months that the river becomes the very spine of the region and large numbers of animals take up residence along the banks.

The river enters Mala Mala in the north-western corner of the reserve, runs eastward toward the hippo pools and then turns south to run past the main camp. In winter when the river is low and game viewing is best along its banks, many points of the river can easily be crossed. A favourite crossing is at the Flockfield Boma, because in the winter this is leopard country.

It is difficult to pinpoint why certain species of animal prefer some areas to others. Perhaps here it is the thick vegetation in the riverbed, the nearby confluence points of the Mshabene and Matshapiri rivers with the Sand, or that it is a popular watering spot for ungulates. It is probably for a combination of these reasons that this has been the core territory of numerous leopards over the years.

At the time of writing, the incumbent male is an easily recognisable young animal. Known as Shica, he features in Gerald Hinde's book, *Leopard*. For a number of months we frequently found him lying in broad daylight on the bank of the river, his eyes closed and his tail hanging in the water and this became known as his fishing spot.

The Flockfield Boma is a large reed enclosure on the western bank of the river above the crossing where the landowners in the Sabie Sand Game Reserve meet for their annual discussion on conservation issues. Many are members of families that have have owned land in the bushveld for years and the names at the meeting today are the same as have been heard by the old jackal berry trees standing over the boma since it was built.

The area around the boma is full of surprises: nyala, though uncommon elsewhere, feed on fallen leaves around the boma and groups of elephant use the track past the boma to get to the crossing. Some of the biggest riverine trees in the reserve occur here: sausage trees with their large inedible sausage fruits, jackal berries tall, thick-stemmed with delicate shiny green berries which are favoured by monkeys, baboons and flocks of fruit-eating birds, matumi trees whose huge trunks are used elsewhere in Africa to make dugout canoes, and the ubiquitous marula, so productive that every year thousands of tons of the soft-fleshed berries fall to the ground in the bushveld providing an important food source for many species of bird and animal.

Towards the end of the summer of 1991/92 it was obvious that the whole of the south-eastern part of the African continent was heading for its worst drought ever recorded. The heat of the summer was intense, the rain which had started with heavy downpours early in the season ceased, the skies cleared and the land became parched. At Mala Mala all the natural waterholes and pans that rely on a good soaking in summer in order to hold water through the early winter dried and grass stocks became dangerously low.

It is at times like these that the two basic resources upon which animals depend for survival, food and water, become further and further apart. The areas that still have water suffer from overgrazing and trampling by concentrations of animals coming down to drink and every day the herds have to move further from water to find grazing. Some animals such as the highly successful impala can deal with the early phase of drought cycles by switching their feeding from grazing to browsing and for this reason impala still look sleek and healthy in the early winter months. The movement to and from water and feeding areas has a number of consequences. The animals spend more energy on travel and therefore need to assimilate more food. They have to traverse areas they would normally not utilise and they spend less time concentrating on their early warning defence systems. In addition, many females in the herds are pregnant and all food is in short supply – their condition weakens and they become easy prey for the ever-vigilant predators. The big cats are experts at the detection and exploitation of weakness, it is their livelihood, and drought presents them with plenty of opportunity. Late in May 1992, the beginning of winter, we were sitting in the landrover watching herds of impala, wildebeest and zebra coming down to drink at the last pools of muddy water in the river. There was a single lioness lying in the shade of a bush beneath an overhanging bank of sand. We watched her sleep for some time. The impala moved towards the water. Two or three took halting steps towards the water's edge only to skitter back to the safety of the herd when dust blew in an eddy around their legs. The herd stood in the lacy shade of three meagre acacia trees, heads moving in all directions with nostrils flared, eyes straining, ears twitching. Nervousness rippled through the herd like a shoal of frightened fish, and again a handful of youngsters edged towards the water.

The lioness now lay on her stomach, ears up, body tensed and head forward, pale yellow eyes fixed unblinkingly on the prey. She was perfectly positioned in a downwind position, staring intently. By now the whole herd of impala were at the water's edge, evenly spaced, and with spread legs and long graceful necks outstretched, they slaked their thirst. We looked again at the lioness, but were stunned to see just empty sand.

She had begun to move under cover of the bank into a position flanking the herd. The wildebeest and zebra began moving in behind the impala, their hooves kicking stones. Low grunts and snorts were carried to us by the wind that blew sand against the landrover. Then the waterhole erupted. Leaping, twisting forms of impala and flying hooves of galloping zebra were lost in a pall of dust as the lioness charged. She flew in, muscle jumping and bulging under thin skin as her powerful limbs propelled her into the maelstrom and then we lost sight of her. We heard no bellow or alarm call but as the dust blew away we saw the lioness again, clinging to her prey.

She had caught an adult wildebeest bull; she bit savagely into his snout, adjusting her grip to sink her long claws into his neck. And so started a dance of death in the dry, dusty heat of a bushveld morning. The wildebeest kicked with his front legs trying to dislodge his attacker. He charged, shoving the lioness down the bank. He swung his head violently, banging her against a granite outcrop at the edge of the sand. She clung onto him with extraordinary strength for twenty minutes. Then, with a titanic effort, she twisted the bull off his feet and we again lost the struggling animals in dust. When it cleared, the bull was back on his feet with the lioness still trying to cut off his breath in a kiss of death. Eventually she brought him down again and the wildebeest succumbed quietly and with some dignity.

The lioness was exhausted. She left her prey and went to rest under the bushes at the edge of the waterhole. After forty minutes her breathing had returned to normal and she moved away from the kill calling in a low mournful "awou" – we realised for the first time that there must be other lions nearby. She returned with another adult lioness and two cubs. They must have been within earshot of the kill but even with their keen hearing had not known of the savage struggle taking place nearby.

In an area where rainfall is erratic and unpredictable, a perennial river like the Sand is all important to the well-being of the ecosystem. Like many of the rivers that run through the Kruger National Park and eventually drain into the Indian Ocean, the Sand arises to the

A martial eagle drinks and bathes in the river. With the largest wingspan amongst African eagles, the martial lives on a diet of game birds, waterfowl, small mammals and large monitor lizards.

A female kudu's large ears act as an early warning system while passing through the thick reeds on the way to the river.

west on the Drakensberg escarpment. This means that before the water reaches Mala Mala, it travels through highly populated areas, industrial towns and farming land. Being at the end of the river's journey places the conservation area at great risk as it is vulnerable to events upstream. In winter, when the agricultural and industrial demand for water is high, the river level drops alarmingly, even after reasonable rainfall years. In dry years the demand for water increases and the reduced flow, the increased silt load in the water, high concentrations of dissolved nutrients and minerals and industrial and agricultural pollution combine to disasterous effect.

In the drought of 1983 the Olifants River in the Kruger National Park became clogged with silt. Fish died by the ton and hippos and crocodiles were stranded in pools of sludge. Eighty years of conservation were lost within hours. Without some habitat management, particularly river management, beyond the borders of conservation areas this will always be a threat.

It is always with great reluctance that I leave my place under the great sycamore and turn my attention from the river.

White tails held high, a pack of wild dogs have some fun with a herd of buffalo.

A hippo bull surges powerfully into the river. The land speed of these huge beasts is quite astounding.

An unusual creamy-backed form of the bateleur eagle bathes in the Sand River. Like many other bare-faced birds, excitement and overheating leads to flushing of the skin.

Overleaf: During the mating season young male waterbuck are ejected from the herd by the dominant male and they spend the next few months with other non-breeding males in bachelor herds.

A pair of saddlebilled storks hunt together on a rocky stretch of the river. Their diet consists mainly of fish, crustaceans and frogs.

The pelage of the mature nyala bull makes it one of the most attractive of all the antelope and helps impress both rival males and potential mates.

The yawning display of a large hippo bull serves to show off his hardware and discourage attacks by rivals.

Right: *The elegant beauty of the great white egret.*

Far right: *A majestic pair of African fish eagles rests on a regular hunting perch while ringing out their familiar call.*

Death of a wildebeest ... the lioness clamps her jaws on the bull's muzzle. He pushes her savagely down a steep bank, but she retains her grip. With her back against the rocks, the lioness tightens her hold and finally wrestles her prey to the ground. After a short rest she calls in the pride.

A yellowbilled hornbill alights on a marula tree above the entrance to its nest hole before delivering a tasty grasshopper to the chicks sealed inside.

The yellowbilled hornbill is one of the most recognisable birds in the southern African lowveld because of its large banana-like bill, black and white plumage and comical ways. It feeds mostly on the ground, searching for insects under leaf litter, in long grass and even by flicking and picking at large piles of elephant dung. Its clucking call, a distinctive bushveld sound, is often accompanied by a spectacular dancing and dipping display with wings held high.

The most interesting part of the hornbill's life cycle is the unusual way in which the adults raise their young. In the summer months the couple will seek out a natural nesting hole in the trunk or branch of a large tree. After mating the female enters the hole and lays three to four eggs. The male brings mud and packing material to the female and she seals the entrance to the hole from the inside, leaving only a small longitudinal slit. She then moults all her primary and tail feathers and in this semi-naked state begins to brood the eggs. The male flies backwards and forwards ferrying insects to his bride; when the young hatch his work load becomes exhausting. When the youngsters are half grown, the mother breaks down the mud and emerges with newly grown flight feathers to help her frantic and exhausted mate feed their perpetually hungry, squabbling brood. The youngsters reseal the entrance. The nest never becomes soiled because the birds take it in turn to aim their rear ends out of the hole to defecate, the whitewash that this leaves on the ground is a good clue to the whereabouts of a nest.

The advantage that sealing the hole confers on the hornbill is of course protection from nest-robbing snakes, birds and mammals such as monkeys, baboons and genets. The danger is that if something happens to the male, the flightless female and her helpless brood are doomed to a lingering death from starvation.

THE GRASSLANDS

A hissing snarl from a young cheetah is a warning to back off.

The Mlowathi Dam is a temporary body of water that fills from a small catchment area on the banks of the Mlowathi River. The dam fills in the summer and so does a series of smaller pans above and below the dam. The water usually lasts into the first two months of winter and becomes the most important wallowing and watering point for kilometres. There are large open areas to the north and east of the dam which support sizeable populations of the larger herbivores such as wildebeest and zebra. Rhino drink at the dam most evenings and large herds of buffalo often water here while moving between feeding areas; giraffe, steenbok, ostrich, impala and kudu are also frequent visitors. The high density of game attracts the attention of predators and death and drama are regular visitors to the grassland of the Mlowathi. Three prides of lions have common territorial boundaries here and often hunt in the riverbed and on fringes of the open grasslands.

Many sightings of wild dogs have been recorded here and it was these animals that were the instigators of one of the most unusual and exciting wildlife dramas we have ever seen.

Late in the summer of 1991 there had been very little rain at Mala Mala and the plains of the Mlowathi were almost bare, a harbinger of the severe drought that was just beginning. Ironically, dry conditions at this time of year are linked to a higher survival rate among the young of plains game like zebra and wildebeest. Their chief predators, lions and hyenas, tend to target youngsters during the calving season in mid-summer. The structure of the bushveld vegetation is such that open grassland is surrounded by wooded areas with taller grass. In normal rainfall years the predators chase herds from the open plain into the long, rank grass of the woodland. Here the youngsters are slowed down in grass higher than their heads and, once separated from the herd, start giving alarm calls which are easily detected by the hunters. This year, after a dry summer, the grass was uniformly short, and an unusually high proportion of wildebeest calves had survived the critical first months of life and were on their way to subadulthood.

It was one of these young wildebeest that became involved in a life and death struggle with wild dogs. We located a pack of 12 dogs moving quickly towards the dam and followed them in our landrover. On reaching the edge of the open area the dogs spotted the resident herd of wildebeest and slowed down. After a short stalk, slightly crouched with ears pressed back, they accelerated to full chase speed.

The sleek athletic lines of the fastest of all land mammals, the cheetah.

55

It is a spectacular sight to see these, the rarest and most efficient of Africa's hunters, in full flight across an open grassland, ears back, mouths open, necks straining forward and white-tipped tails held stiffly up, in pursuit of their prey.

The dogs soon caught up with the herd and were lost in the dust, snapping at flying hooves in the apparent chaos of the hunt. They separated a young calf from the herd and the lead dog closed in. She clamped her powerful jaws on the young wildebeest's hind leg, pulling it down into a shallow pan of water. Normally this would have been the end of the hunt, but the dogs had not bargained on the courage of the calf's mother. She dashed back and scattered the surprised dogs giving her calf time to struggle to its feet. The dogs were not fooled for long. They regrouped and formed a threatening ring around their prey, once again they pulled the youngster down. We were certain that the drama was over, when there was an amazing turn of events. A herd of zebra which had been standing some distance from the fracas trotted up to the wild dogs and began lashing out at them with well-aimed kicks, braying and biting viciously at members of the pack. The dogs were again startled into retreat and this time the mother and her badly shaken, but only slightly wounded, calf made good their escape back to the safety of the herd.

One is tempted to interpret this as altruistic behaviour, which is not normally thought to occur in nature; for one species to help another does not benefit the genetic survival of members of their own species. In this case it may have been a defence response by the zebra, although adult wildebeest and zebra are not generally attacked by wild dogs in this part of Africa, and this herd of zebra had no foals themselves.

Some prey species do help one another in the shared struggle against predators, whether by actively mobbing them and chasing them off, or by sharing watchfulness. Mixed herds of giraffe, baboon, impala, zebra and wildebeest are commonly seen. All of these animals have strengths that complement one another in predator detection. Giraffe can see over nearby cover, baboons have excellent eyesight both from ground and treetop level and impala, zebra and wildebeest hear and smell predators at a good distance. In this way advantage is conferred on all by teaming up.

The plains of the Mlowathi also provide a perfect hunting habitat for the fastest and most graceful of the African cats, the cheetah. Cheetahs are still the rarest of the big cats and in fact have only recently been taken off the red data book list of endangered species; they are now categorised as threatened. More often than not cheetah are found lying or sitting in an elevated position; in the bushveld this is usually on a termite mound or fallen tree trunk. There is an air of nervous tension about these animals. When lying on their sides, their heads are always cocked up and constantly moving from side to side, orange eyes searching and tail tips flicking. When standing, the cheetah displays a beautiful silhouette, the deep chest, long thin legs, narrow waist and small streamlined head are a result of perfect fine-tuning and adaptation for speed hunting. The long upswept tail has a laterally flattened end that acts as a counterbalance and rudder for high-speed turns, while the spinal column is as flexible and supple as the young green branch of a bushwillow tree.

The claws of the cheetah are permanently unsheathed to provide good traction for high speed sprinting.

Although not adept climbers, cheetah will often clamber onto low branches to get a better view of the surrounding country.

The cheetah is a rather enigmatic animal with a fascinating history and prehistory. The cheetah's historical distribution was throughout Africa, most of India, Afghanistan and the Middle East, in fact the name itself is derived from the Hindi word "chitah" meaning "spotted". Very small numbers of the Asian subspecies *Acinonyx jubatus_venaticus* still remained in Kazakhstan and Iran before the political instability in these counties; their status is now unknown. In Africa the healthiest populations of cheetah are in the southern subcontinent and the savannas of East Africa. At some time in the prehistory of the cheetah's evolution, the species suffered what is known as an ecological "crunch" period, when unfavourable conditions or perhaps an epidemic led to a drastic reduction in cheetah numbers. Possibly only a few individuals or maybe even an individual pregnant female remained. This had the effect of reducing the gene pool dramatically, which is evident in the extreme similarity in the gene make-up of the world population of cheetahs. The huge disadvantage which this confers on the species is that there is not the genetic diversity to make some individuals stronger or more resistant than others,

and as a result cheetahs are extremely susceptible to epidemics brought on by viruses or bacteria. This may have a bearing on the low numbers of cheetah in Africa. Other explanations for the naturally low numbers of cheetah are its hunting method and its frailty as a result of its adaptation to speed.

Cheetahs are very obvious predators, they always sit on elevated viewing sites and when hunting use very little stealth in pursuit of prey. As a result the herd animals which they hunt are all aware if a kill is made and very soon get "spooked" if a cheetah hunts the same area for more than two or three days. Cheetahs must therefore always move on, in order to do this their territories are very large – up to 500 square kilometres. Even allowing for a good deal of territorial overlap, which does occur, they spread themselves fairly thinly, resulting in low population numbers.

Cheetahs have sacrificed strength for mobility, they use velocity to bowl over their prey rather than strong shoulders and neck to wrestle and kill larger animals. A large heavy skull with deep-rooted teeth for holding and throttling large prey is replaced by a high-domed, short-muzzled skull with large nasal passages so that the cheetah can breathe while holding onto the throat of an already downed and

In the first two months of life young wildebeest are vulnerable to attack by a range of predators. For this reason they stay close to their mothers and the protection of the herd.

A short crouching stalk is followed by an explosive burst of speed.

exhausted prey animal. This lack of strength and their largely solitary way of life places them at the bottom of the large predator hierarchy, below the social power of wild dogs, lions and hyenas and the arrogant singular strength of the leopard. All these animals will rob cheetah of prey or actively expel them from hunting grounds; the specialist is surpassed by the generalists.

In late 1989 rangers at Mala Mala began sighting what they thought to be a king cheetah, the animal was extremely shy and could not be approached closer than 200 metres so the sighting was not confirmed for some time. The king cheetah is a genetic variant of the species, in much the same way as the black panther is a melanistic form of the Asian leopard or the South American jaguar. The king cheetah possesses an allele (part of a chromosome) that has an aberration in a particular spot known as the tabby locus. This spot contains the genetic code for coat patterning and is only present in the cheetah populations in the Eastern Transvaal in South Africa, the south of Zimbabwe and the north-east of Botswana. The gene is recessive and manifests itself very rarely in the wild, but can be bred in captivity. Much success in this regard has been achieved at the De

59

Wildt breeding centre in South Africa. The result is a splendid animal with large oblong and heart-shaped chocolate brown spots and dark stripes running the length of the back and tail. The tear marks that run from the inside corner of the eyes to the black-edged lips of all cheetahs are especially pronounced in the king.

Sightings of king cheetah in the wild are rare and only one other had been sighted in the previous five years. Early in 1990 the sighting at Mala Mala was confirmed. A magnificent female king cheetah was photographed near the Matshapiri River. This was enough reason for celebration, but there was an added bonus. The female had three young cubs – one of which was a female also bearing the king cheetah markings. This family began to appear fairly regularly between the Mlowathi and Matshapiri rivers. With careful attention to the comfort of the animals we gradually habituated them to vehicles, eventually reducing their flight distance to less than 15 metres.

When the cubs were 16 months old their mother suddenly left them to fend for themselves. The subadult king female suffered a leg wound, apparently from a hyena bite, and began to limp badly. Ordinarily this would spell certain death for a single cheetah. We alerted the vets in the Kruger National Park immediately and a decision contrary to our normal non-interference policy was taken. They would attempt to capture, cure and then reintroduce the animal. The king cheetah's rarity and the fact that she was still part of a post-independence sibling hunting group influenced the decision positively. She was tranquillised, captured and taken to Skukuza for medical attention. After 10 days she was deemed fit to return to the wild but it was important to first locate her siblings before her release, as her chances of survival would be at least 70 per cent lower if she was released alone. Everyone was on the lookout, but it was another eight days before the young male and female were spotted near the Nkapene River. A radio call was immediately sent through to Skukuza where a team was on standby for the translocation.

The situation seemed perfect, the two youngsters were on a large impala kill and were likely to be there for some time. The crate holding the young cheetah was loaded at Skukuza and reached Mala Mala an hour later. The door was opened and the cheetah stepped delicately out before bounding off into the bush. We followed her for most of the day as she tried in vain to rejoin her siblings, calling in a high bird-like chirp. This strange call is an adaptation to the cheetah's low position on the predator hierarchy ladder; it is seldom detected by competitors as a predator call.

Fears that the king cheetah would be rejected by her siblings were allayed the following day when we found all three feeding on a freshly killed duiker. The release was the first of its kind involving a king cheetah; its result vindicated our decision to interfere with nature.

In the spring of 1992 rangers at Mala Mala sighted the young king cheetah female with two tiny spiky-haired cubs, both fit and healthy although neither displayed the king markings.

The status of cheetah in Africa is still critical – in 1955 there were an estimated 28 000 animals, in the early 1970s a census recorded only 14 000. The population has grown since then, but habitat destruction and their position in the predator hierarchy will keep their numbers low.

A wild dog rests after an early morning kill; these hyperactive animals will sometimes hunt from dawn till dusk without pause.

The African buffalo, *Syncerus caffer,* has the reputation amongst hunters of being the most formidable of the Big Five African trophy animals. This reputation is built on their bulk and strength, on their resistance to injury and on their intelligence, described by Theodore Roosevelt as "brutish nastiness".

In some parts of Africa, particularly along the richly grassed foodplains of the great rivers, buffalo travel in herds of up to 2 500, but in the bushveld of the Sabie Sand area herds have seldom numbered more than 600. It is estimated that only 1 in every 10 000 buffalo survived the rinderpest epidemic of the 1890s. By 1910 their numbers had normalised, a testament to their resilience and reproductive success.

Large herds of buffalo are the most important bulk grazers on the plains because they remove the long coarse grasses, exposing the lower new shoots for more specific grazers. Buffalo have very stiff immobile lips, so grazing is effected by a long prehensile tongue which wraps around a bunch of grass, pulling it into the mouth and chopping it off between a wide row of incisors and the upper palate with a sideways movement of the head.

The social organisation of the buffalo is interesting to watch in large herds. People often ask: "Who leads the herd, who starts them moving?" It has long been assumed that buffalo society is matriarchal, but in fact it is the prime reproductive bulls who are dominant in the herd. The leaders of movement are individuals that are often not dominant or even adult. These are known as pathfinders and seem to take the lead in turns. The rest of the herd consists of subgroups or clans of related cows and their offspring from the last three years; these groups are attended by a number of subadult or adult males. There is a direct linear hierarchy amongst

the cows and a set hierarchy amongst the bulls. When the herd is stationary and grazing the subgroups spread out but stay together. When resting the buffalo lie down to ruminate, clans will lie together in tight groups with each individual touching the one next to it. It is while moving, particularly through wooded country, that the herd is most vulnerable to predators, especially lion, and for this reason a specific formation is used when travelling. The pathfinders move out in front attended by a small number of prime herd bulls. Behind them come the bulk of the herd made up of three or four subgroups travelling in tight formation. On the flanks of these groups are bachelor males and in the rear are weak or injured individuals that are unable to move quickly, followed by three or four prime herd bulls as a rearguard. If danger threatens the pathfinders quickly drop back and the dominant cows and bulls encircle the herd facing outwards to protect the young in the centre of the formation. This formidable sight is usually enough to discourage attack, but if the assailants persist they are charged by a number of the biggest herd bulls. This group protection allows badly injured and weaker animals time to recover, even blind buffalo have been seen to enjoy good health in herds for years.

It is the older males that have left the herd to form bachelor herds or to live as loners that have perpetuated the buffalo's reputation as a mean and fearful adversary. These bulls are called "dagga boys" by the Shangaan trackers, a reference to the *dagga,* mud, with which they cake themselves to keep insects off their hides which become more vulnerable to parasites as they lose hair in old age. These groups are fairly sedentary, living in a small home range usually near a river where they lie on the cool sand during the heat of the day, moving onto the grassy banks in the

evening to graze. The older these animals get, the more short tempered and aggressive they become. This might result from their vulnerability to predation in small groups and because older animals carry a higher load of irritating parasites, especially during the summer months.

The herds of buffalo at Mala Mala

follow a fairly regular drinking regime during the winter months usually watering once or twice a day in the Sand River. It is then that their herd organisation seems to break down. The movement toward the river is orderly until the first animals smell water, when the lowing and bellowing begins and the youngsters and subadults break rank and gallop towards the river. Soon the whole herd is stampeding into the water, splashing and sucking noisily. In the drought years when their condition is low, this behaviour has often proved fatal. Lions take advantage of the chaos and attack animals which they normally would not risk challenging.

The rare king cheetah.

The long grey mane on this cheetah cub's back will disappear as it reaches adulthood. It serves as camouflage in long grass and is even thought to simulate the appearance of the ferocious honeybadger, a good deterrent to other predators.

Previous page: *A young waterbuck bull in a misty winter landscape.*

The Mlowathi Dam is surrounded by grasslands.

The square-lipped or white rhinoceros has bounced back from the brink of extinction in southern Africa.

67

A determined team of wild dogs attack a young wildebeest calf, but are turned back by a courageous mother.

The pack again grabs the hapless youngster and this time death seems certain. But intervention by a herd of zebra allows another miraculous escape and the bloodied youngster races off with his mother to the safety of the herd.

A redbilled oxpecker calls from the horns of a young buffalo bull.

Not many days go by in the bushveld of which the forktailed drongo does not become a part. The Shangaan people believe that if the drongos call in the morning it will be windy, and they are usually right. Drongos are insect eaters that have developed a commensal relationship with most of the larger game species of Africa. They follow herds of antelope, zebra, buffalo and giraffe or small groups of rhino and elephant hoping to catch the insects that are disturbed by their passing. For this method of hunting to be effective suitable perches are essential as vantage points. As a result the drongos hunt mostly along the ecotone between grassland and woodland. This way they can perch close to the animal, occasionally letting out a strident chirping to let other drongos know that the space is occupied. As soon as an insect is flushed, most often a grasshopper or beetle, they swoop down, catch it on the wing or in the grass, and fly back to their perch to enjoy the meal. The drongo's beater has usually moved on a short way by the time the meal is over, so the

drongo follows, flying in a shallow arc to the next close perch.

As with most adapted behaviour, there is a reason for this hunting strategy. Firstly, the presence of animals ensures a higher chance of well camouflaged insects being disturbed and becoming visible. And secondly, ensuring you have a perch means energy will not be wasted in flight while searching for prey. The result is a very energy-efficient hunting method and an extremely successful species. The question that begs to be asked of this story is "Why does the drongo not sit on the animal?" It seems that the change to perching on moving animals is a big evolutionary step: foot structure has to change, and this would affect its ability to perch on thin twigs where the drongo carries out the other important phase of its life cycle – breeding. The answer for some drongos is to perch on horns occasionally and leave the clambering over backs and legs to another insect specialist, the oxpecker.

Redbilled oxpeckers, like drongos, are very noticeable in the bushveld. They have developed thickened tail shafts for propping against an animal's body and have opposing pairs of toes to grasp onto the host's skin. This condition, known as zygodactylism, is seen in other climbing birds such as parrots and woodpeckers. Oxpeckers play an important role for their hosts, cleaning the larvae of potentially harmful parasites out of the hair and skin. One oxpecker can eat up to 4 000 ticks in larval form in a day. It is not surprising then that oxpeckers are tolerated by most animals and can often be seen burrowing into ears in search of a tasty morsel or pushing their beaks into noses for a drink of nutrient-rich mucus. Apart from relieving their host of the irritation of being bitten by parasites, the oxpecker serves as an avian prophylactic, because in Africa ticks and other ectoparasites act as vectors for an

Forktailed drongo.

enormous number of viruses and bacteria-carrying diseases that can debilitate or kill the animal on whose blood they feed.

Occasionally when an animal is injured in a fight or scratched by a branch the oxpecker can turn from an ally to an irritating free-loader. The large scars that are often seen on the necks of giraffes or the saddle of a buffalo's back are testament to this swing in the relationship. Oxpeckers peck at the flesh of an open wound, keeping it open. This allows flies to lay eggs in the wound – the oxpeckers then feed on the hatching larvae and the wound can take excessively long to heal, sometimes resulting in the death of the animal.

There is an interesting story behind the distribution of oxpeckers in the Mala Mala area. Of the two southern African species, only the redbilled has occurred here until very recently. The extinction of the yellowbilled oxpecker locally was ascribed to agricultural pesticides used to dip livestock against ectoparasites. The yellowbilled oxpecker may have had a lower resistance than the redbilled, and their numbers did not recover after the mass die-off of bovids in the dreaded rinderpest epidemic of the late 1890s. A project was initiated in 1990 to re-establish the yellowbilled oxpecker in parts of its original range and it was with great excitement that rangers at Mala Mala began recording these birds among a large herd of buffalo in late 1991.

LIONS: THE PRIDE OF AFRICA

One of the Styx Pride lionesses scratches her back on a wild date palm in the river bed.

The supremely confident gaze of the monarch.

There is no animal symbol as important or as evocative in human history as the large-maned lion. Ancient drawings and sculptures are rich with lion images and numerous important buildings, bridges and castles are guarded by sculptures of lions. This fascination with lions was probably fuelled by fear and respect for a powerful fellow predator, but more recently scientific research has encouraged a renewed fascination as our understanding of lions has grown.

SOCIABILITY

The lioness is always the focus of studies on communal living in lions. A pride usually has a nucleus of two or more females and can consist of up to eight or nine related females: mothers, sisters, cousins, aunts or daughters. The females and their cubs form a stable coalition that raises cubs, hunts and defends a territory together. The advantages of a family grouping such as this are best explained in terms of the theory of kin selection. A lioness's own cubs carry 50 per cent of her genetic make-up and 50 per cent of their father's genes. If her cubs survive to adulthood the lioness has been successful in fulfilling her genetic potential. If in turn she suckles her daughter's cubs, she is helping to ensure the survival of 50 per cent of her daughter's genes – which are effectively 25 per cent of her own genes – further increasing her contribution to the future. The genes that code for this behaviour are likely to become more frequent in the population with each successive generation showing this behaviour. In addition, by helping another female's cubs a lioness is also ensuring that her own cubs will have companions to share pride duties when they grow older.

Males do not show such a deep commitment to their kin; unrelated males will often form coalitions. A male lion's chances of survival on his own are slim. He is not as versatile a hunter as the female, his bigger bulk makes him slower and his mane makes him more obvious on the hunt. A male on his own is very likely to fall foul of territorial males during the nomadic phase of his life. A single lion has very little chance in a fight with a coalition of two, three or four.

When young males are evicted from prides they need to scavenge and hunt together – and whether the group is just a gathering of unrelated youngsters or a true brotherhood split off from their natal pride, their chances of survival are enhanced by their sociability. Sometimes a group of young males may only leave their pride at three or four years of age when their manes are well grown and they are large and powerful. A group such as this is likely to march into a neighbouring pride and take over, as their size and large manes will scare rivals from a distance. Once established they take up the pride duties of protection and territorial defence – roaring, scent marking and chasing off strangers. Studies have shown that members of large male groups sire more offspring than members of small male groups, and are therefore genetically more successful.

It is very rarely that even the most sophisticated research projects are able to follow and map out the full lives of succeeding generations of large mammals. In any African game reserve the major attraction is always the predators, and the lion, being the largest and strongest of them, receives the most attention. For this reason rangers at Mala Mala have been following and documenting lion behaviour for many years and a wealth of fascinating data on pride structure, lineages, longevity and reproduction has been gathered.

THE STYX PRIDE

The Styx Pride is a fairly typical lion pride in Mala Mala although it has consistently had low numbers due to a poor cub survival rate. It does however demonstrate many of the interesting facets of lions' social behaviour. At the core of the pride are three females – two sisters and a daughter who, along with her brother, is the only survivor of a litter born late in 1989. Their range is huge and they hold territory in the north-eastern and central part of the reserve where they overlap with the ranges of a number of other prides. In late August of 1991 the lead female disappeared to give birth. The following is an excerpt from Gerald's field note book:

4 September 1991

The pug marks of the Styx Pride were easy to follow as they headed south along the open sand of the riverbed. They swung east out of the river, through the thick reeds, past stands of wild date palms and onto the riverbank. We tracked on more cautiously, as the visibility was limited in the thick bush by long grass under the large, shady riverine trees. My tracker, Hendrik, held up his hand cautioning me to stop. He listened for a few seconds and we moved on. It was early morning and the dew-drenched grass made tracking easier. The long grass bent in the direction the lions were moving, as we crossed over firmer, rocky terrain. Hendrik's keen eye and tracking experience picked out clues of an animal's passing that seemed invisible until pointed out – an overturned stone, a leaf full of sand, a broken twig or a toe print on flat ground.

Ahead of us lay a heavily wooded donga known as the "ngoboswane". We suspected this may be where the lions were resting up, as for many years it has provided shelter for prides of lions during the day. Sure enough, as we entered the thicket on the lip of the donga we heard a soft, deep growl. It is a sound that one never gets used to – it comes rushing back from the beginning of time and warns you as it did the early hunters of the world. It automatically freezes you to the spot. As our eyes grew more used to the shady darkness, we could make out the form of a lioness lying facing us with her head low and her ears flattened against her massive skull. The low rumble continued, rising slightly in pitch with each exhala-

The dominant Clarendon male gazes wistfully at a leopard kill cached out of his reach in the branches of a marula tree.

Night falls quickly in Africa, and lions begin the evening hunt.

tion. Adrenaline made our arms grow heavy and stomachs hollow. We backed off slowly, but the movement caused the growling to rise and intensify as it does preceding a charge. We stopped and stood dead still. After a few seconds, which passed like an hour, we started back again and gradually moved behind the bank of the donga. We hurried back to the landrover and after some manoeuvring got ourselves closer to the lioness. She was still there, but very much more relaxed with the presence of the vehicle than she had been with the two humans who had disturbed her. As I was setting up my cameras Hendrik touched me on the shoulder and pointed toward the rear of the lioness. A curious little face was peering at us through some short grass.

Oversized ears, sparkling eyes and wrinkled brow gave the little cub a gnome-like look, an irresistible little ball of fur. As the shutter clicked, the face disappeared and we saw a number of little bodies scurrying for shelter behind mother's bulk. With the curiosity which characterises all small cats, it wasn't long before these newcomers had to get their second look at a landrover, and four identical little faces came up one by one over the lioness's back. Soon the cubs clambered over their mother and, with some muted mewing and gurgling, settled down to suckle. We could make out the rest of the pride lying flat out and disinterested on the far side of the donga. This was our first sighting of the new cubs in the Styx Pride.

The two males who had fathered the Styx Pride cubs were first seen in the Clarendon Dam area in February 1991 and became known as the Clarendon males. When they first made contact with the Styx Pride they were very wary, obviously unsure of whether there were other males in the area. However they soon established themselves by ousting an alliance of two adult and three subadult males, known as the Borehole Pride, who had been nomadic in the area for some time. We first saw the Clarendon males mating with two young females from the Msagwene Pride. By now they were becoming more relaxed around the landrover and were sighted more often. In April they mated with the Styx Pride for the first time. The females did not conceive and it is likely that they were just getting to know the new males, as is often the case after a takeover.

But in due course a lioness of the Styx Pride did conceive and after Gerald's first sighting the cubs became a regular attraction on game drives. It soon became obvious that the Styx females were not going to mollycoddle their four new additions and it was not long before the cubs were accompanying the pride on hunting forays. It is extremely dangerous for young cubs to be taken along on the hunt − it exposes them to other predators, they may be injured at a kill or get separated and lost. In addition if the hunt is unsuccessful, as often happens with lions, the cubs will have expended an enormous amount of energy just to keep up with the pride and for growing youngsters suckling is not enough to replenish that energy.

By the time the cubs were six months old they looked only half that age because of poor nutrition. Ironically they were born at a time when food supply was abundant in the form of young impala, zebra, wildebeest and warthogs. This abundance affected them adversely because the females would kill small prey animals, just enough to satisfy their own hunger, and then stop hunting. Sometimes while following them we saw the cubs go without food for two or three days. Their condition deteriorated steadily.

The turning point came for the cubs when the pride moved to the top of the koppies across the river from the camp and began hunting the open grasslands where there was a high population of large ungulates − wildebeest, kudu and zebra appeared on the menu more regularly. The cubs' condition improved markedly. It was here that a second lioness in the pride gave birth to four cubs who soon joined the hunt. The pride now consisted of three adult lionesses, four youngsters, four small cubs and a subadult male, the brother of the younger female, who had stayed with the pride despite vicious attacks by the Clarendon males.

But then tragedy struck. The older cubs were moving with the pride one night when they encountered two strange males on the Manyolethi River. A pride fight ensued which lasted some three hours − the earth-shattering roars and bellows of male lions kept us awake. The next morning we found the mutilated bodies of two cubs on the riverbank and only one cub remained in the pride. The smaller cubs were also destined to a short life, they disappeared when they were about six weeks old.

In stark contrast to the poor mothering behaviour displayed by the Styx Pride a lone lioness we came to know well, the Nkapene female, proved to be an extremely successful mother and taught us not to

76

Lionesses are serenely patient with their constantly active cubs.

The Styx Pride females.

How do you do?

generalise about these fascinating beasts. A steady territory with ample food and the security of two successful males would have led one to predict that the Styx Pride would be more successful at raising young than a lone female in an unfamiliar territory without the patronage of males or companions to hunt with. The reverse was true.

THE NKAPENE FEMALE

The most fascinating individual amongst the hundreds of lions that have been members of resident prides, nomads or temporary residents in Mala Mala, has undoubtably been the huge lioness who came to be known as the Nkapene female. She was originally a member of a large pride that held territory in the north-eastern part of the reserve. She was particularly recognisable because of her size and very light colour.

When I first saw her she was the dominant female in the Msagwene Pride with another lioness and five youngsters ranging between 9 and 14 months. I learned that her first two offspring were the dominant females of the Mlowathi Pride who had seven cubs between them. It was intriguing that this lioness, in the prime of her life, was a grandmother with a second litter approaching independence. She had the reputation of never having lost a cub to natural causes – a reputation she kept until the very end of her tenure at Mala Mala.

It was an evening in the crisp, cold winter of 1989 when I first saw her skill as a hunter. Her pride were hunting on the northern side of some koppies on a large stretch of bare red earth which never seems to grass over, even in the wettest of summers. We had been following their spoor for most of the afternoon and evening when the spotlight at last picked out a long line of eyes in the road moving straight towards us. John Sibuye, our tracker, was delighted and whispered excitedly "*ngonyama*", lion – 16 years of tracking cannot dull the thrill of locating a predator. We pulled off the track and watched the lions walking towards us – legs swinging, large paws slapping forward, honey coloured bodies moving fluidly, shoulders working up and down in a casual display of power. There is something almost hypnotic in watching a pride of lions on the hunt at night.

The big blond female stopped next to the landrover and lifted her head to let the breeze bring information to her. She slowly turned her heavy head and neck to stare through us with pale golden eyes. It is a look that takes you back to your primordial roots, centuries of evolution and years of experience and knowledge fall away under that stare. She swung away and continued up the track with her hunting partner and family trailing in her wake.

In the weak glow of the new moon we saw the lions freeze as they picked up a movement on the fringe of the koppies. It was a herd of zebra crossing towards the dry riverbed on the northern side of the clearing. The large female sank into a low crouch and slowly began to move towards the herd. She broke into a trot and disappeared from our view. We sat silently for a short while before we heard an explosion of hooves as the zebra broke into a gallop. I started the engine and raced into the cloud of red dust towards the sounds of scuffling and grunting. At a steep bank we saw the Nkapene female at the throat of an adult zebra.

The cubs were at an age when they were learning how to deal with

Evening activity begins with a bout of yawning and stretching.

live prey and their mother obviously thought this an opportune moment to test their new skills. The cubs gathered around the zebra trying various holds. Their mother let go, and for some 30 minutes we witnessed the grisly struggle until the zebra succumbed to its awful wounds.

Only a few days after this we found the Nkapene female alone in the Sand River. With a suddenness that was hard to understand she had left her pride. When they made contact the next night she was extremely aggressive, mauling one of her cubs and biting her ex-hunting partner. The reason for this behaviour became evident eight days later when we found her mating with a large black-maned lion who had established dominance in the northern territory. The noisy and tumultuous love affair lasted for three days during which time the lions mated evcery 10 to 30 minutes.

The Nkapene female made it quite clear that, having served his purpose, the male was no longer needed and he beat a weary retreat. Her unprecedented behaviour baffled and disturbed us, our celebrity lioness had abandoned her pride and her territory. Two weeks later we found her some 10 kilometres from her home area investigating holes in termite mounds along the banks of the Nkapene River. From then on she lived and hunted completely alone.

Her behaviour changed in a way that seemed to indicate that she had an innate knowledge of the most successful strategy for living safely and raising the cubs that were growing inside her. She became a solitary hunter, changing her prey selection and hunting tactics accordingly. She targeted smaller prey and put maximum effort into an early kill. Warthogs became her favoured food. She would begin hunting at dusk just before the warthog families were settled in their burrows and became adept at flushing them and swiftly despatching them as they emerged.

In early September of 1990 her cubs were born, and although rangers knew where she was holed up it was some time before they were seen. She moved back to her old territory and gave birth in the ngoboswane donga, to the east of the Sand River and about a kilometre from the camp. Eventually she brought her cubs out to introduce them to a landrover. She walked towards the vehicle, lay down very close to the front wheel and began calling with a low mournful "Oooh". One by one, five tiny bow-legged cubs staggered towards her, spitting and mewing at the large unfamiliar green object near their mother.

The Nkapene female spent two weeks in the donga and then over a period of weeks moved the den slowly southwards to the Nkapene River, the territory where she spent her pregnancy. Such was the efficiency of her hunting that she would kill almost every night. Unlike females in other prides she never took her cubs with her but left them safely hidden at the den and after killing – no matter how far off – would return to lead them to their meal. On a number of occasions we observed her kill and rest for a while before returning to her cubs. If on the way back she came across more game she would kill again. One night we saw her kill three warthogs from the same burrow. They ran in separate directions on emerging and were at an extreme disadvantage in the dark, she methodically hunted down each one of them. She then carried them to a central point where she made a pile of prey. On her way back to the den she killed a

The Styx Pride take an early morning rest in the river after an unsuccessful night's hunting.

The Nkapene female – an incredible athlete and a superb mother.

young impala ram. She fetched the cubs and returned to the warthogs, ignoring the impala kill. But she had not forgotten the fourth kill, by morning the lion family had finished off all three warthogs and the impala.

As the cubs grew, they spent more and more time with their mother on the hunt. She started capturing large prey and letting the youngsters get involved in the kill. On one occasion she caught a young giraffe and let the cubs practise dispatching it – an act that looked cruel, but was an important learning experience for the cubs. When her cubs were eight or nine months old the wise old lioness gathered her family and moved into the Kruger National Park, they were never seen again. Stories of the big blond female lion are still told by rangers, trackers and guests around the campfire at Mala Mala.

While in Mala Mala the Nkapene female did not spend any time marking or defending a territory, she seemed to want to remain anonymous and concentrate her energies on raising her family. This

was probably just as well because her close neighbours were the Dudley Pride, who often hunted the Nkapene area.

The Nkapene cubs enjoy an evening meal under their mother's watchful eye.

THE DUDLEY PRIDE

We first identified the Dudley Pride in the autumn of 1989 – an elderly lioness with two adult daughters. In spite of her advanced years it seemed that the Dudley pride mother had lost none of her enthusiasm for the pleasures of the flesh, she was seen mating regularly with the incumbent males of the north known as the Buffalo Boys. Later that year all three females were seen mating with these two magnificent males – this marked the beginning of a year of wonderful sightings of the Dudley Pride.

On a summer's evening in 1990 I was on my way back to the camp with some visitors when a large lioness suddenly stepped out of the dense bush onto the road. She was followed by two more females who were quite obviously hunting. My guests were tremendously excited by the prospect of a hunt and I prepared to follow them, but John Sibuya whistled softly and cautioned me to wait. A tiny lion cub swaggered into the road in his mother's footsteps. One by one, in a long line, came his brothers and sisters. There were ten cubs in all – the largest common litter I have ever seen. We did not need to wait long before the females killed an impala and the little ones joined them in a noisy feast. It was a beautiful sight and the first of many that the Dudley pride were to treat us over the next year.

As the months went by the old female became weaker and weaker, partially losing her sight and lagging behind at kills. Then one

Only when the cubs have eaten well will she join the feast.

evening we saw the Dudley pride but she was missing. She was never seen again. The hunting behaviour of the pride changed as soon as the old female disappeared. We first noticed it when one of the rangers found a freshly killed kudu – uneaten and still warm. We arrived as he found the spoor of two lionesses heading south along the river. This was Dudley Pride territory, but their practice had always been for the females to eat their fill and then for one of them to go and fetch the cubs if there was meat leftover. However without the old female to compete for the kill, leaving the carcass untouched would have a number of advantages. Other predators would be less likely to detect the kill in the absence of feeding noises or the smell of an open body cavity. They would not lose meat to the old female while they were fetching their cubs, and they could bring the rapidly growing cubs to an intact animal which is invaluable experience to teach them how to handle prey.

Sure enough the lionesses were soon located – calling softly for their cubs. After two or three minutes their calls were returned with the sharp, excited "eeoow" of the cubs, who came barrelling out of the bush, jumping, licking and biting their mothers in a frenzy of excitement. The females stood and accepted this rough and noisy adulation with the resigned patience of mothers who know that their children will never give them a moment's rest. They led the cubs back to the kill where the cubs moved their attention from the females and settled down to feed.

As the cubs grew the demand for meat increased and the Dudley females began to seek out larger and larger prey which eventually led to the death of the lead female and the demise of the pride. They had been hunting in the south of the reserve and were found one morning on a fresh giraffe kill. On approaching the kill we noticed the tail and hind paws of a lion protruding from under the huge male giraffe. After a quick head count we realised that they belonged to the dominant lioness − she must have sprung onto the giraffe's side and brought it down on top of herself. She probably died instantly. The pride fed on the kill for ten days before pulling the female's body out from under the carcass, but they did not eat her flesh. After the death of the lead female the pride disintegrated. The youngsters continued to hunt in the area for a while before splitting up and moving off. We were saddened by the death of an animal we had watched for years, but felt privileged to witness another fascinating act in the ongoing drama of nature.

All the large African predators are, to a greater or lesser extent, opportunistic. The cheetah is perhaps the most specialised of all and their prey selection in terms of size and species remains fairly constant and predictable, lions on the other end of the scale have specialised in being generalists. The king of the beasts has reached his lofty position and achieved his success through some very humble behaviour. Lions have been recorded taking prey such as snakes, flying ants, tortoises, nestling birds and mice. But in the winter of 1990 we recorded a kill on the other end of the scale.

Lion cubs often play in the low branches of trees.

Late one evening I was walking to my room when I heard a high-pitched wailing from across the river. It sounded quite a long way off, and vaguely familiar. The almost human wails continued at short intervals and I wondered if some unfortunate person had been treed by lions. Suddenly there was a louder squeal and a low explosive snarl, immediately recognisable as that of a lion and I realised what the other sound was. I had last heard it when watching two bull rhinos fighting some years before. At the time I had been struck by the incongruity of that thin scream coming from such a huge beast. The scenario became clear. About a month before a young bull rhino had taken up residence on the floodplain opposite the camp. He was at an age when he should still have been running with his mother or a family group. The Styx Pride had been hunting in the area over the last few days and, unlikely as it seemed, they were attempting to kill the young rhino. We set off to investigate immediately. As we rounded a corner the headlights picked up the glint of eyes and then the heart-rending sight of the young rhino sinking down onto his haunches and rolling over onto his side with the three lionesses and young male from the Styx Pride clinging to him. He lived for another five minutes before dying with a final sad, whining expulsion of breath.

It was evident that the fight had lasted for at least half an hour and the lions may well have been circling and tormenting the rhino for more than an hour and a half. There were no wounds on the rhino that looked substantial enough to have been fatal. Indeed once the lions began feeding it took a full hour before they could break the skin around the ear to get at the flesh. It appeared that the helpless young animal must eventually have given in to the stress of the attack. Certainly this is one of the very few recorded instances of rhino being attacked and killed by lions.

84

The Nkapene cubs play in the early morning on the track to their den.

Predators hunting in the same area usually influence one another negatively, but occasionally predator species can benefit from one another's hunting strategy. Gerald and I witnessed just such a series of events in the Nkapene female's territory. We came across a family of cheetah on a very fresh impala kill. The female had four five-month-old cubs who were all settling down to feed. As we stopped to watch them we heard a duiker distress call from around a bend in the river about 150 metres from the kill. A duiker darted across in front of the vehicle and jinked quickly along the riverbed. The area had been recently burnt so the female cheetah saw the fleeing duiker and immediately gave chase. This was extremely unusual behaviour for a cheetah that had just spent an enormous amount of energy bringing down an impala ram. To our great surprise she caught the duiker after a 200 metre chase and dragged it under a nearby bush where she was joined by her cubs. The impala was left untouched. After a little while we drove down the riverbed and found the Nkapene female and her large cubs feeding off a baby duiker. Her kill had obviously flushed the female duiker towards the fortunate cheetah.

Right: *An affectionate cub greets the Dudley Pride female.*

The Nkapene female toys with a baby duiker killed minutes before its mother was caught by a cheetah upstream in the Nkapene River.

The Dudley Pride cubs settle down to feed on a kudu bull to which their mothers have brought them.

The sad end of the Dudley Pride matriarch, crushed to death under her last kill, a huge male giraffe.

Drinking at the Sand River.

Looking up at annoying monkeys.

The mutilated body of a lion cub

The driving forces behind infanticide are the importance of ensuring genetic representation in the next generation, and the instinct to wipe out the competition for resources within hunting grounds. Competition for food is intense among Africa's predators, and hyena, wild dog, cheetah, lion and leopard will all kill one another's young if they get the chance.

Within species infanticide also occurs, although it is most common among the social predators, especially lions. Lion cubs have a long maternal dependency period of up to two-and-a-half years. During this time their mother will not come into oestrus as hormones stimulated by the presence of cubs suppress the ovulation cycle. Male lions usually have a short period of dominance in an

A mother leopard carrying the half-eaten remains of her dead cub.

area before being ousted by new males. If the females in the pride have young cubs a newly dominant male will have a long wait before mating. However if they kill the cubs immediately the females will come back into oestrus within a month. This greatly increases the chance that the males will still be dominant while their own cubs are born and raised and thus carry their genetic heritage through to the next generation. Male cubs that are over a year old when new males join a pride are considered a potential threat to dominance and are driven off, whereas older female cubs are accepted as potential mates.

Lion cubs can also be killed by females from rival prides or may die from wounds received during fights for meat at a kill.

Newly dominant male lions may kill cubs in order to bring their mother into oestrus more quickly.

A number of instances of male lions killing cubs have been recorded at Mala Mala. In 1990 we watched two nomadic males kill seven of eight cubs in the Charleston Pride; an interaction that lasted a full day with the females desperately trying to protect their young. The surviving cub was ignored during further contacts with the males until it disappeared at five months old. A more recent incident took place when two six-month-old cubs from the Styx Pride were killed by interloping males. The two dominant males in the territory fought off the interlopers but in the morning we found the badly-mutilated bodies of the cubs on the banks of the Manyolethi River.

Wild dogs, lion and cheetah very seldom eat the remains of cubs they have killed, but hyenas and leopard do so frequently.

THE ELUSIVE LEOPARD

An afternoon rest.

Powerful curved claws help make the leopard an accomplished climber.

The beauty and grace of the cheetah are matched by the other spotted cat of Mala Mala – the leopard – who exchanges a sleek athletic frame for a lithe power-packed body, open grasslands for dense riverine bush and day for night.

The leopard is the most adaptable and hence the most successful of the large hunting cats. It enjoys a distribution throughout sub-Saharan Africa, along the North African coast into the Middle East, Asia Minor, South and South East Asia and into parts of the Far East. The secret of the leopard's success is its ability to live and hunt in areas outside of protection, "in the shadow of man" is how Dr John Seidensticker describes it in *The Great Cats*. He quotes the example of a radio-collared leopard which researchers tracked entering a village in Nepal one night. The leopard spent the night trying to catch a goat from the penned herd in the centre of the village, but had had no luck by daybreak. Finding its avenue of escape cut off the cat spent the day tucked away in a woodpile while the village activities went on around it, not even the village dogs knew that it was there. After dark that night it strolled out of its hiding place and disappeared into the forest.

There is nothing anatomical or morphological that makes the leopard stand out from other big cats. It does not have the swiftness of the cheetah, the size of the lion or tiger, the thick coat of the snow leopard or the stocky legs and large canines of the jaguar. This may well be its secret: the leopard is a generalist with the advantage of behavioural flexibility. It can change its hunting habits to survive in habitats from desert to thick forest, from remote areas to the suburbs of African towns. It preys on an amazing array of species and will switch to whichever prey source represents the best return for energy expended. Another advantage that leopards have over other large cats is that they cache their kills in trees out of reach of man and other predators.

Home range sizes vary enormously, depending on the productivity of the environment. The largest home ranges recorded were for two male leopards in the Kalahari Desert that had home ranges of 4 000 square kilometres. In the mountains around Stellenbosch in the Cape, a radio-collared male ranged over 487 square kilometres. At Mala Mala the home ranges are between 15 and 60 square kilometres with an average for females of about 20 and 50 for males.

The leopard's solitary and secretive lifestyle has until very recently made it an enigma. A few years ago one could spend long periods in the bush and would consider oneself lucky to get a glimpse of a spotted blur as it disappeared into a thicket. Man had always been the leopard's primary enemy, coveting its magnificent coat. To survive the onslaught of poachers and the retribution of angry stock farmers, leopards had to become masters of stealth and concealment. By early 1970 conservation groups succeeded in making it unfashionable to wear spotted cat skins. The International Fur Trade Federation called for a halt in the endangered species fur trade and in 1975 CITES (The Convention on International Trade in Endangered Species of Wild Fauna and Flora) listed the cheetah, tiger, jaguar, clouded leopard, snow leopard and leopard on Appendix 1 prohibiting international commercial trade in these species. Today leopards are threatened by loss of suitable habitat more than by hunting. Because man no longer persecutes leopards they have become less secretive and in areas such as Mala Mala have even become more diurnal. Few places in the world can compete with the leopard-viewing we have had at Mala Mala.

"Nyow-nhyou, nyou-nyou, nyou ..." the urgent alarm call of the vervet monkeys drifted towards us from the river. The infectious alarm call that vervets reserve for ground predators alerted the entire troop; they jumped from branch to branch trying to get a better view of the leopard who made his way hastily into the thick undergrowth. How many times in the past had we welcomed this and other alarm calls knowing that they will lead us to a predator – very often a leopard.

On our arrival in the area the alarm calls were of moderate intensity indicating that the danger had moved off. The few monkeys still giving the intermittent alarm calls were perched high up in a strangler fig, looking upstream. We made our way along the track that hugs the lip of the embankment before swinging away to climb steeply towards a lookout point. Lying comfortably along the branch of a sausage tree was the leopard we had been searching for; his head still and his gaze piercing as he stared in the direction of the monkeys. As the alarm calls subsided he closed his eyes, his solid limbs hung loose and his head gradually sank until his chin was firmly stretched along the branch. Every time a sound dragged him from his half-sleep, an ear flickered and heavy-lidded eyes opened wearily. The last rays of the setting sun glowed amber on the elegant, dangling tail.

A grey duiker moved hastily across a clearing and the leopard, attracted by the rustling of dry grass, lifted his head, his keen eyes following the movement. He hesitated for only a moment before hurrying down the tree, moving swiftly along the ground and through the donga that separated the two animals. Only the occasional trilling of a small scops owl broke the eerie silence of early evening. The leopard cautiously approached the top of the donga. The duiker was moving quickly west towards the airstrip, the leopard followed, but his movements became less intense as he lost sight of the duiker.

It was June of 1990 and we were collecting the final photographs and data for the book, *Leopard.* The leopard we were following was the nineteen-month-old Mlowathi male who had gained independence from his mother in this area some five months before. Because he had grown up in the presence of landrovers he was relaxed and paid no attention to us if we kept the required distance

Beauty, grace and power are blended in the leopard as in no other animal.

The pugmark of a male leopard in the damp sand of a bushveld game path.

from him. We spent a lot of time with the Mlowathi male during his post-independence period and shared many of his learning experiences – this night was one of the highlights.

Reaching the edge of the open ground around the airstrip the Mlowathi male paused, making sure that there were neither lion nor hyena in the vicinity, then proceeded cautiously across the open ground into Princess Alice's bush. The leopard gave little more than a glance towards a large herd of wildebeest grazing in the safety of the open area. As we followed into the heavily wooded area a large male warthog broke cover in front of our vehicle. The leopard gave chase, caught up with it and vaulted onto its back. With the warthog's head firmly between his jaws, a battle began that lasted a full 25 minutes. The leopard eventually managed to move his grip from the head to the throat and squeeze the fight out of his powerful prey. Sapped of energy, the leopard rested next to his hard-won meal.

The battle had been long and noisy. Long because in his inexperience the Mlowathi male had taken on prey that was too large for him; he was lucky to escape unscathed against such a formidable opponent. Noisy because the warthog squealed continuously throughout the fight. After a few minutes' rest the Mlowathi male suddenly sat up, stared into the darkness and fled. A pride of nine lions came into the light of our spotlight to claim the kill.

The ostrich is the largest of all living birds. Its size prevents it from flying but its powerful legs generate high speed to outrun predators.

The Mlowathi male had learnt a valuable lesson. Although the size of the prey was worth the energy expended to kill it, the prey was too big to drag into a tree. The noise of the struggle was bound to attract stronger predators such as lion or hyena and they would be sure to expropriate the meat. Lessons such as these must be learned quickly and without injury in a leopard's post-independence period. Leopards are loners and cannot afford an injury that would render them incapable of hunting, unlike lions that have a social structure. If a lion is temporarily injured the rest of the pride will continue hunting and allow the injured animal to feed on the leftovers.

The next day we found the Mlowathi male walking upstream in the reedbeds of the Sand River. He was investigating opportunities for a meal, scampering after cane rats and sniffing amongst the reeds. Suddenly he froze, then stalked forward. He had seen a small bushbuck lamb hiding in the reeds. The lamb jumped to its feet and ran up a sandbank in the middle of the river with the leopard in hot pursuit. The Mlowathi male clipped at the little buck's long legs, snagging it with one claw, playing it like an angler would a fish. Seeing the opportunity for a game he released the buck which, in an obvious state of shock, slowly moved away from him. He inserted a claw in the soft skin and pulled it back. A cat and mouse game ensued. The lamb tried to cross the river, but the leopard gingerly entered the water to retrieve his meal. After 45 minutes of play, the leopard despatched the terrified creature with a bite to the head. He hoisted the body into a large marula tree near the flood plain, tossing it as he jumped from branch to branch. In his subadult period the Mwolathi male was always playful. In the years to come this side of his nature would disappear as he settled down to the serious business of survival.

For the next few weeks we watched him regularly inspecting warthog burrows in the late evenings as the numerous warthog families were returning for the night. One evening he flushed an adult

The Mlowathi male, a beautiful young male leopard whose core territory was in the Flockfield Boma area.

female and her three eight-month-old youngsters from their burrow. He crouched beside the entrance, the muscles of his hindquarters bunched beneath him as he anticipated the warthogs' moment of exit. He could hear the shuffling as the family became restless and moved towards the mouth of the burrow. The sow was wary, sensing danger, she broke cover from the termite mound suddenly, her youngsters following in single file, with tails erect. As the last piglet emerged, the leopard burst forward, uncoiling like a spring, he bowled the piglet over and grasped it in his jaws. The sow heard the squeal and came rushing back to rescue her hapless youngster. The leopard dropped the wounded piglet and bounded off, keeping his distance from the sow's flashing tusks. The warthog family escaped down a burrow some 30 metres off. Unperturbed by his dangerously close encounter, the leopard set up station next to the new retreat. For a second time the warthog family bolted, sow first then the piglets. Once again the leopard waited for the last piglet, the same one he had wounded in the first encounter. Bounding forward he bowled it over with his paw, and bit into its neck. The sow returned, this time making contact with the leopard before he dropped the piglet. The leopard, the sow and the piglets scattered in all directions. The Mlowathi male chased after the badly wounded piglet narrowly missing it as it scampered down a third burrow. The mother and the other two piglets disappeared across the road. The leopard started a long and patient vigil, carefully entering the burrow every now and then to try and retrieve his meal. It was not until 22 hours later that he made his final lunge into the dark hole, and appeared with the piglet feebly struggling in his mouth.

Male leopards are usually forced, by the local territorial male, to become nomadic when they reach sexual maturity at between two and a half and three years of age. Circumstances were different for the Mlowathi male as although his movement was slightly erratic over the years to follow, he kept popping up in the same area. As he grew, his head broadened and he acquired the thickset, muscular physique of an adult male. His playfulness fell away and he became more serious. When we tracked him on foot as a subadult, he always ran from us, stopping about 15 metres away and looking back at us curiously, almost playfully. As time passed he was less likely to run but would lie staring at you with pale, murderous eyes.

One day in September 1992, when the Mlowathi male was 34 months old, we came across him in the company of his mother. Her behaviour was typical of a female in oestrus. She slunk back and forth in front of him, rubbing her hindquarters in his face, then crouched in front of him, inviting him to mount her. At first he did not accept the invitation, greeting her advances with deep growls. She repeated her performance several times before he mounted her and we heard the stuttering outbursts as he climaxed. The mating continued over a period of three days, but as far as we know no cubs were conceived. The Mlowathi male tarried in this area for a few more months and then disappeared. When we found the badly decomposed carcass of a leopard near the Sand River we presumed that he had finally paid the price for his defiance of the territorial male. But this was not the case as he reappeared in the vicinity a few weeks later.

A leopard and her large cub gambol in the early morning on their way to a waterhole.

Overleaf: Leopards spend most of their time in dense bush and will only cross open country when absolutely necessary.

Our day's work in the bush during our period of intensive study on leopards would start long before dawn. We would drive sleepily along the bumpy roads watching the eastern horizon turn from deep navy blue through pink and crimson before fading into the cold blue of daylight as the sun rose. Winter mornings are freezing cold. The wind bit through to the bone as we travelled in the open landrover. Drowsiness would be replaced by expectation as we travelled through secretive places that had become familiar to us. Crossing the Sand River we warmed our aching bones in the first rays of the morning sun.

On one such morning we scanned the yellow expanses of the sandy riverbed as the sun came up and noticed a movement some way off. Through binoculars we saw a leopard struggling with an antelope. We drove quickly towards the action and arrived to see a hyena rush down the embankment and chase the leopard off the now dead bushbuck ram. The hyena ripped into the soft underside of the groin, swallowing great chunks of flesh. The leopard watched from the top of the embankment until the hyena had finished a third of the carcass and then rushed down in an attempt to chase off the hyena. But the hyena repulsed the attack and dragged the kill to the top of the embankment, ate his fill and left. The leopard, who became known as the airstrip female, returned to the remains of her meal.

99

The Mlowathi male drinking from the Sand River.

That night, when we returned to camp Gerald compared the sketch he had made of the airstrip female's facial markings with those of a young cub photographed for his book, *The Original Gamedrive*. They were identical. She was the surviving cub from a litter of two; her sibling was killed by lions in December 1988 at about three months old. It was almost two and a half years since we had seen her and we were excited at the prospect of re-establishing ties with her. We had been watching her mother since April 1982 and continued to do so until she disappeared in 1991, when she must have been 15 or 16 years old, although it has been commonly believed that leopards seldom live beyond 12 or 13 years in the wild.

The airstrip female has established her home range adjacent to that of her mother, where fortuitously there was a territorial vacuum when she became independent. Her home range included a few kilometres along Sand River, which is preferred leopard habitat and home to many prey species. The reedbeds were a favourite haunt for the airstrip female during the first year and a half of post-independence, they offered adequate cover and an abundance of small prey. She proved, as did her mother, to be a good hunter and adept at self-preservation, but as with many leopards, inexperience led to the death of her first litter of cubs.

Late one night in June 1990 we were travelling back to camp along the river road from the hippo pools when we saw a leopard walking away from us with what looked, at first glance, like a cub in her mouth. On closer inspection we recognised the leopard as the Marthly female and saw that she was carrying a scrub hare. Her loose belly skin and the matted hair around her teats indicated that she was lactating.

For two weeks we searched the area between Manyolethi River and the hippo pools on the Sand River, desperately trying to locate the Marthly female and her cubs. Eventually we gave up the continuous search and resigned ourselves to the fact that she had disappeared into a secret world. Leopards are masters of this art and countless numbers live undetected from one year to the next.

Then, a month later, we came across fresh leopard spoor at the junction of the Manyolethi and Mlowathi rivers – a female leopard and two cubs. The search resumed. It was very early morning and we tracked cautiously on foot; a leopard with cubs can be extremely aggressive. Two years previously we had had a close call with the Marthly female when she had two six-week-old cubs. We were following her spoor for a short distance from the landrover when suddenly a low-slung yellow blur came flying out from behind a termite mound. The tracker let off a shot which sent particles of sand and stone flying into the leopard's face and, fortunately for us, the leopard ran off into the bush. With this experience in mind, we tracked for only a short distance on foot, returning to bring the vehicle closer every now and then. We stopped the vehicle in the Manyolethi riverbed and walked towards a rocky outcrop. The spoor was extremely fresh. It is crucial for the tracker to know how fresh the spoor is and to look out for other signs. Although it was still early the leopard's tracks were superimposed on top of the fresh tracks of a diurnal animal and urine was still dripping from the leaf of a date palm where she had marked territory. We felt uneasy and decided to return to the vehicle for an early morning cup of coffee while we waited to see if the leopards and cubs would appear. While we were pouring the coffee Hendrik, the tracker, wandered off towards the base of the rocky outcrop and as we turned to call him we saw him running towards the landrover with the leopard in full charge 15 metres behind him. There is no animal more intrinsically savage than a leopard intent on attack.

It is generally believed that if one stands one's ground in the face of a predator charge the animal will not follow through the attack. This time Hendrik did not stay to test the theory. He ran for the landrover clambering onto the tracker's seat with impressive agility considering that he was weighed down by a heavy duffel coat. We shouted loudly and the leopard stopped. We waited some 15 minutes to allow her, and us, to settle down and then followed her. The Marthly female lay out in the open with two little spotted grey balls of fur who quickly disappeared into the thick raisin bush behind their mother when we arrived. She snarled as we approached. We sat some distance away and waited for two hours before the first of the cubs ventured to the fringes of the bush to peer at the landrover.

Over the next two months we saw more of the Marthly female and the cubs, but they always remained somewhat elusive. When the cubs were five months old we found their mother roaming her territory

The delicate blue waxbill is one of the prettiest of the guild of bushveld seed-eating finches.

calling at intervals, her distress evident. This behaviour continued for two days and nights. When she eventually returned to an island in the Sand riverbed, in her core territory, she had only one cub with her. Two months later the second cub disappeared. In seven years she had not yet brought one cub to independence.

During their adulthood female leopards are always rearing cubs or pregnant. Leopard mortality is high, particularly in areas of high lion density. The younger and more inexperienced the female, the higher the mortality rate of her cubs is likely to be. In the Mala Mala area leopard cub mortality is estimated to be as high as 65 per cent. Towards the end of our leopard study period at Mala Mala, the Marthly female once again disappeared and it was not until Gerald's last day in the area that we saw her again.

We decided to drive around the territories of the four leopard that we had come to know best to catch a final glimpse of the animals that had so captivated us. We left camp at sunrise and set off along the western boundary of the reserve crossing the junction of the Sand, Mlowathi and Manyolethi rivers – entering the Marthly female's home range. We found her at the junction moving towards the thick reedbeds of the Sand River. We watched as she disappeared once again into her shadowy and secretive world. The same morning we found the airstrip female in a sausage tree and the Mlowathi male near the tamboti thicket. Gerald's farewell was complete except that we found no sign of the Flockfield male, the dominant male leopard who provided us with valuable information.

An almost grown leopard cub looking out for the return of its mother.

Powerful shoulder and neck muscles, sharp curved claws and phenomenal strength allow the leopard to cache its prey – a distinct feeding advantage.

104

Clinging on, the leopard subdues his powerful quarry. The Mlowathi male tackles a huge male warthog who tosses him over his back. After about half an hour of struggle, the exhausted leopard quickly leaves the scene as a pride of lions, attracted by the sound of the battle, take over the kill.

Once prey is safely cached a leopard can rest between bouts of feeding.

Previous page: *The Mlowathi male with his hardwon prey. He spent 22 hours trying to extract this warthog from its burrow.*

The airstrip female on territorial patrol.

An unusual sequence of photographs shows a female leopard soliciting a male by flirting her tail around his face. He mounts her briefly, biting her neck as the mating intensity increases. When it is finished, the female turns savagely on the male, beating him with her front paws.

A huge lappetfaced vulture dominates smaller whitebacked vultures at a zebra carcass.

The sight of vultures sitting on a large dead tree, shoulders hunched against the setting sun, is one if the most powerful and evocative images of Africa. In the lowveld there are five species of commonly occurring vultures, forming what is known as a guild of species. Although despised for their habits and initially repulsive looks, vultures are fascinating birds with a very important function as the dustmen of the bush. Interesting behavioural and anatomical differences allow them to co-exist while seeming to compete for the same resource. The first vulture to arrive at a carcass is usually the small, thin-beaked hooded vulture, who is often alerted to the presence of food by the champion carrion finder, the bateleur. Soon the sky begins to fill with dots as more and more of the airborne cleaning crew come to investigate. The social feeders, the whitebacked and Cape

vultures begin to flap noisily down into the trees around the carcass and onto the ground some distance away. After inspection, one or two of birds will flap heavily over to the carcass and begin feeding. Suddenly the seemingly disinterested birds on the periphery fly and run towards the carrion and a scrum of cackling, squealing, flapping and bickering will ensue with as many as 200 birds competing for scraps of meat. The hooded vulture does not have the heavy bill and long unfeathered neck of the social vultures, nor does he attend in large numbers. He usually feeds on scraps gleaned from the outskirts of the feast which he can pick up in his delicate bill. It is not long before the king of the African vultures arrives. The huge seven kilogram lappetfaced vulture falls from the sky like a malevolent archangel on wings spreading nine feet from tip to tip. Usually only one of two

of these magnificent raptors attend a feed, but their huge size and aggression ensures their dominance. If a carcass is untouched it is only the massive beak of the lappetfaced vulture that can break through the skin. When the bones are picked clean, this bird can continue feeding on skin and small bones long after his fellow feasters have left.

The fifth species is the enigmatic, solitary whiteheaded vulture whose diet seems to consist more of smaller carrion such as mongooses and squirrels pirated from other raptors. At the large carcasses only one or two of these vultures may appear. They seldom get involved in the violent and undignified pushing and shoving among the rotting flesh.

Early winter is breeding time for all these species and eggs are usually laid between May and July. The whitebacked vultures breed colonially in the tops of trees and it is remarkable how many colonies are found in the core area of lion prides.

  Because of their large size and wing span, the bigger species have to wait until late in the morning for the earth to warm and for thermal columns of hot air to start rising before they can spread their wings and gain height for their search for food. The smaller vultures do not rely on thermal soaring and are active early in the morning which explains their early arrival at the remains of a night hunters' feasts.

Hooded vulture.

A CONSERVATION SUCCESS STORY

The sable unit at the southern end of the Main Camp.

The word "conservation" is often misinterpreted. Conservation is just one of a number of wildlife or natural system management options. It involves protecting a specific resource or system of resources from change. In southern Africa wildlife management policy is increasingly turning to a wise use of sustainable natural resources as an alternative to static preservation policies. The basic philosophy behind these policies is that animals, plants and environments are best protected if they are worth something to the humans living around them. If a habitat can be maintained and cared for so that wildlife flourishes, and money is generated as a result, then a strong argument can be made for perpetuating such a system. Many die-hard nature lovers are shocked that nature, wildlife and animals should be "exploited" to generate money, but it is this money that will win over the local population's interest, and allow finances to be poured back into habitat maintenance programmes. This improves natural game populations and attracts tourists who provide the funds for further improvement. It is through a commitment to a sensible use of sustainable resources that Mala Mala has been able to maintain and improve its corner of Africa.

Mala Mala's land management policies have an interesting history. In the late nineteenth and early twentieth century hunters decimated the vast herds of game that grazed on the sweet grasslands of the bushveld until many species became locally extinct and numbers dropped alarmingly. In 1920 a decision was taken to raise cattle in the area that now constitutes the Sabie Sand Game Reserve, including Mala Mala. This represented a short-term threat to indigenous animals in the form of competition for grazing for herbivores and direct extermination of predators that threatened the livestock. The long-term effects are still being felt 70 years after cattle farmers gave up their unequal struggle against disease and predation.

Cattle are adapted to life in moderate climates, and only a few breeds such as the Zebu-derived cattle of India, are even slightly adapted to low water and sparse grazing conditions. Few breeds are resistant to the diseases of Africa and no cattle have grazing patterns that fit into African ecosystems. The grasslands of Africa are extremely delicately balanced systems. They are also the most productive of any system, save perhaps the oceans, in terms of species diversity and animal numbers. These species can only co-exist if they all fit into specialised trophic niches in the pattern of grazing. Each species utilises a different part of the grass plant at different times of year.

A hole dug in the Sand River bed by elephants brings sweet underground water filtering up from below the surface.

117

This is known as temporal or spatial niche separation. Rhino and zebra may graze the rank, coarse grasses seeding in late summer; wildebeest graze on the lower portion; impala, elephant and waterbuck graze low shoots and sweet springtime grass, steenbok and duiker eat the small forbs; and tsessebe and sable select only their favourite plant species from among the grasses. Thus a huge biomass of disease resistant, low water consuming animals are supported.

Cattle, sheep and goats graze all levels and types of grass on the plains, removing the binding roots and leaving the topsoil vulnerable to erosion by rainfall. To slake the thirst of cattle, which require up to five times as much water as indigenous species, dams were built which concentrated the trampling effect of the herds around water and caused further erosion. Roads were built to get to areas where cattle were being grazed, and were placed along the paths of least resistance, through the grasslands, which caused still more erosion.

When cattle ranching came to an end at Mala Mala, an era of hunting began with very little or no habitat management taking place, apart from some haphazard burning programmes. Important scientific research and an enlightened approach to the task of restoring the bush to its natural condition began in the early 1970s. The pioneering work in the Transvaal lowveld was carried out by Ken Tinley, who described the functioning of seeplines.

A seepline operates at the junction of the sandy crown of a low bushveld hill and a sill of clay on the slope of a valley. Water penetrates the loose sand but as it cannot filter through the clay, it seeps out to run down the valley floor. During the rainy season the low-lying areas become waterlogged providing ideal conditions for the germination and growth of grass species. The seeds of trees and shrubs cannot germinate in waterlogged areas so the grassland is dominant. With bad management these delicate areas are easily damaged. Overgrazing and trampling by cattle and bad placement of roads compact the soil and diminish its water-retaining characteristics. Water runs off the compacted soil forming drainage dongas and the lower watertable allows trees and shrubs to germinate. As the larger plants grow they draw more and more water from the soil, accelerating the change from grassland to woodland. Seepline management aims to reverse this process by taking out the "straws" that suck up the water, allowing more water to become available and creating the waterlogged conditions in which grass thrives to maintain the character of grasslands.

The first step in rejuvenating a seepline is a controlled burn. This removes debris and some of the younger woody growth. The next step is to chop down all the young woody growth encroaching on the grassland below the seepline leaving only the larger trees intact. The chopped-off woody growth is then used to pack the drainage lines at the lower end of the seepline to prevent further erosion. Roads are moved to above the seepline.

One of the greatest enemies of habitat management is soil erosion. Erosion processes are always insidiously at work. Water moves over bare earth, wind blows away precious top soil. Over-use by man and animal leaves large patches of bare earth vulnerable to these elements, which must be nursed back to health.

Natural systems maintain themselves in a state of dynamic equilibrium; there are always changes, cycles of ups and downs. Mala

118

The incomparable beauty of an African sunset.

Mala is in a marginal rainfall area with an average annual precipitation of 700 millimetres a year, most of which falls between October and April. Drought is a regular phenomenon, a severe one occur roughly every 10 years.

In 1979, rainfall dropped after four years of good rain and the reserve was caught in the grip of a searingly hot, dry period that lasted until 1984. Many animals died – especially the water-dependent impala, kudu, warthog and buffalo. The game's condition weakened and there were easy pickings for the predators who culled the slower, weaker animals. The lack of grazing in the critical winter months meant that herd females were undernourished and either failed to conceive or did not carry foetuses full term. Lambing and calving rates dropped alarmingly and predators and scavengers could not keep up with the supply of carcasses. The hot evening breezes carried the smell of decaying flesh; everywhere was depressingly still. As worrying as it seemed at the time, the bushveld and its inhabitants are resilient. As soon as the rains came the recovery process started and within two years the vegetation had regrown and the animals began to breed again. At the end of the dry period there were only 50 skeletal buffalo left in the reserve, but five years later there were herds totalling about a thousand. Impala, kudu and warthog population numbers followed similar patterns.

The dry years produce excellent predator-viewing and sightings of lion, leopard, cheetah and wild dog are more frequent. The autumn and winter of 1990 was the beginning of another dry cycle which offered us a unique opportunity to research and photograph leopard.

Severe drought, often attended by other climatic changes such as extreme cold or hot spells, is cataclysmic and one cannot rely entirely on the resilience of the habitat to re-establish vegetation and population levels. A sound wildlife management philosophy is needed to determine what measures should be taken to alleviate the effects of drought. To what extent can one intervene? When the drought phase beginning in 1992 was predicted, it was decided to initiate an intensive culling operation on key species within the reserve. The rationale behind the decision was that there are a small number of species that have a major impact on available vegetation. When the food resources are critical, these species endanger the survival of rarer animals. In a finite reserve like Mala Mala, the bulk grazers must be monitored, buffalo, rhino and impala being the most important.

These three species are dealt with in different ways. The white rhino has bred with phenomenal success in the Sabie Sand Game Reserve after their introduction at Mala Mala in 1964. When the seriousness of the impending drought became apparent, it was decided to capture and move 75 of these huge animals to alleviate pressure on grazing and to ensure their survival in areas where the drought was not as serious. Rhino capture has been perfected over 35 years and the operation now runs with a smoothness akin to a well orchestrated military manoeuvre. Within three weeks the operation had been completed.

The buffalo posed a different problem. Previously, excess buffalo had been captured and sold, but now there were too many buffalo and no market for them. Buffalo are carriers of foot-and-mouth disease which is extremely dangerous to domestic stock and can therefore not be moved west of the greater Kruger National Park area. All the areas to which buffalo could be moved were suffering under the drought conditions. The decision was made to cull 400 animals from the herd. A culling team from the Kruger National Park were called in and the animals were darted from a helicopter with an overdose of the anaesthetic scoline. The darted animals were removed to the processing factory in Skukuza within three hours.

The third key species, impala, are controlled year round by Mala Mala's own culling team. They are shot at night with a low calibre rifle in the beam of a powerful spotlight. They are immediately taken back to camp where they are skinned and the carcasses dressed and hung in a freezer room. All skins and meat are used on the reserve.

The importance of conservation and protection of the world's natural resources has become increasingly evident over the last 20 years as technological advancement and exponential population growth threaten the very existence of the planet as a functioning ecosystem. Nothing we manufacture and invent can replace or even approach the complex web of interrelated functions of the natural world. The nations more enlightened to this problem are the more highly developed. They have attained their position largely at the expense of the natural environment. The quest for wealth has plundered the earth's resources with a disregard for the fragility and balance in natural ecosystems. These developed nations all have similar trends of reduced population growth, generally high living standards, political stability and functioning civil infrastructures. This stability should provide a platform from which to positively influence developing nations in the Third World. Unfortunately this ideal is not realised. In developing nations maintenance of the environment is very low on the list of priorities which is topped by a struggle for survival amongst the poor peasant majority and a race for riches amongst the affluent minority.

There is no continent as vulnerable as Africa. Almost all her

A controlled fire, part of the late winter burning programme.

120

countries are subject to political instability. A quick succession of governments engender an attitude of "get while the getting is good", and very few long-term development programmes embracing sustainable utilisation of natural resources are initiated. It is not our intention to bewail the fate of wildlife in Africa, but rather to point to an example of sensible conservation management in southern Africa.

The nimble feet and steady balance of the klipspringer.

One of the most delightful of Africa's small antelope is the klipspringer, a stocky little animal with the poise of a ballerina and the agility of a mountain goat.

Klipspringers have adapted to life on rocky outcrops in order to escape predation on the open plains and in the bushveld. Up here they are safe; no animal can match them for nimble speed on smooth granite rocks. This lifestyle has given rise to a number of interesting adaptations in anatomy and behaviour. The most important part of the klipspringer's rock climbing equipment is its feet. One would expect large soft hooves to give good purchase on rock, but instead each leg is tipped by a tiny paired hoof with a soft cup-shaped underside. When standing with all four legs together, a klipsringer's hooves could easily fit onto half a playing card. The klipspringer's coat is also adapted to life on the rocks. Each hair is hollow, like a miniature porcupine quill, and loosely attached to the skin much the same way as a bird's feather. If they rub up against the abrasive

surface of a rock, the hairs cushion the shock and pull out of the skin as the animal rubs past. This prevents the skin from tearing and reduces the risk of breaking a bone.

No other antelope has as pretty a face as the klipspringer. The huge dark eyes are underscored by large pits on either side of the nose. These are the pre-orbital glands and are used by the males in scent-marking their territory. The male will push the pit up against a twig about 60 centimetres from the ground and leave a sweet-smelling, black tarry ball that acts as a warning to intruders. One of the most astounding stories of nature's adaptability surrounds this territorial habit and the life history of a parasite that lives exclusively on the klipspringer. The tick, *Afrixodes matopi,* starts its life cycle as an egg which hatches in the leaf litter and humus of rocky outcrops. When it hatches the larval and nymphal stages attach themselves to hyraxes, rock hares or elephant shrews which are also associated with rocky areas. Now the problem fac-

ing the tick is how to get onto a klipspringer! Ticks usually position themselves on vegetation and wait for any animal to brush past. *Afrixodes* has a number of problems. It is specific only to the klipspringer and the klipspringer is highly mobile, territorial (so only small numbers occur on any one rocky outcrop) and spends most of its time on rocks where there is no vegetation to ease the passage of the tick onto its host. Studies in the Matopos National Park in Zimbabwe have revealed the tick's ingenious secret − it simply climbs the twigs that are regularly marked with the orbital glands of the klipspringer. In the study a survey showed that of the bushes on which ticks were found, 84 per cent were above overhanging rocks and only 16 per cent were in areas where klipspringers were unlikely to go. How do the ticks know which bushes to climb? This remains a mystery, but they are probably attracted to the scent of the territorial markings.

The klipspringer – a stocky animal with the poise of a ballerina.

A subadult female king cheetah is released back into the wild after being nursed by Kruger Park veterinarians.

Previous page: *This baboon seems to be enticing the young impala closer. Baboons will sometimes kill newborn antelope to feed off the milk stomach and tender flesh.*

The delicate and speedy steenbok is the smallest of the antelope species at Mala Mala.

The ugly snare wound around this zebra's neck alerts one to the ever present danger of poaching in African game reserves.

DWARF MONGOOSE:
Team work in the termite mounds

A dwarf mongoose marking its territory.

Looking like tiny brown bears with furry tails, dwarf mongooses dash from refuge to refuge in busy groups of up to 30. The group is led by a dominant female and her mate, known as the alpha pair. These families are thought to be among the most social of all animals and duties such as baby-sitting, standing guard, caring for sick family members and attacking predators are shared. Each dwarf mongoose group forages within a territory marked out in a comical headstanding ceremony in which each member pastes twigs and objects near the termite mounds in which they sleep. The territory has a number of dens, usually termite mounds or hollow logs, in which the family sleeps. It usually takes them about 30 days to do a round of their territory, sleeping in different holes most nights. This constant movement allows food sources to be replenished, new insects to hatch out, and it also seems that parasites like ticks and fleas cannot breed efficiently in the dens when the occupants are away for long periods. The mongooses know their territories intimately and not only know exactly where the dens are but also where to run for temporary refuge if danger threatens.

Danger is part of everyday life for such small but conspicuous creatures and guard duty is an extremely serious business. In the morning when the group moves out onto the western side of the mound to sun themselves and indulge in some social grooming a guard is posted immediately. There is a division of responsibility for sentinel duty with certain male members contributing more. The guard chooses a high position some distance from the group and alerts them to danger with a different high-pitched call for each of their most feared groups of predators, birds of prey, mammals or snakes.

Within the group the mongooses are peaceful but confrontations with other groups do occur in competition for favourite mounds when foraging areas overlap, and these can be quite vicious. The dominance hierarchy within a group is not contested by fighting but by a grooming competition. Whoever grooms the other longest wins. The loser is usually wet and bedraggled, but happy after the bout.

BIBLIOGRAPHY

The works cited have been particularly helpful in the field and in compiling the information in this book.

Attenborough, David (1984), *The Living Planet*, London, Collins.

Branch, Bill (1988), *Field Guide to the Snakes and Other Reptiles of Southern Africa*, Cape Town, Struik.

Estes, Richard Despard (1991), *The Behaviour Guide to African Mammals*, California, University of California Press.

Fabian, Anita (1982), *Transvaal Wild Flowers*, Johannesburg, Macmillan.

Hinde, Gerald (1992), *Leopard*, London, Harper Collins.

Maclean, Gordon (1984), *Roberts' Birds of Southern Africa*, Cape Town, John Voelcker Bird Book Fund.

Main, Micheal (1990), *Zambezi: Journey of a River*, Johannesburg, Southern Book Publishers.

Newman, Kenneth (1992), *Newman's Birds of Southern Africa*, Halfway House, Southern Book Publishers

Schaller, GB (1972), *The Serengeti Lion: A Study of Predator-prey Relations*, Chicago, University of Chicago Press.

Seidensticker, John (1991), *Great Cats: Majestic Creatures of the Wild*, Pennsylvania, Rodale Press.

Scott, Jonathan (1985), *The Leopard's Tale*, London, Elm Tree Books.

Scott, Jonathan (1991), *Painted Wolves*, London, Elm Tree Books.

Scott, Jonathan (1992), *Kingdom of Lions*, South Africa, Russell Friedman.

Skinner, JD & Smithers, RHN (1990), *The Mammals of the Southern African Subregion*, Pretoria, University of Pretoria.

Rattray, Gillian (1986), *To Everything its Season*, Johannesburg, Jonathan Ball Publishers.

Paynter, D & Nussey, W (1992), *Kruger, Portrait of a National Park* Halfway House, Southern Book Publishers.

Photograph acknowledgements
William Taylor: pages 22, 32, 33, 52, 63, 74, 78(top), 82, 83, 90, 126 and 128.
Richard du Toit: pages 40, 41, 44, 77 and 88.
Wayne Hinde: pages 68 and 69.
Nils Kure: pages 25 and 48.

Female impala move away from the herd in early summer to give birth to a single lamb.

INDEX